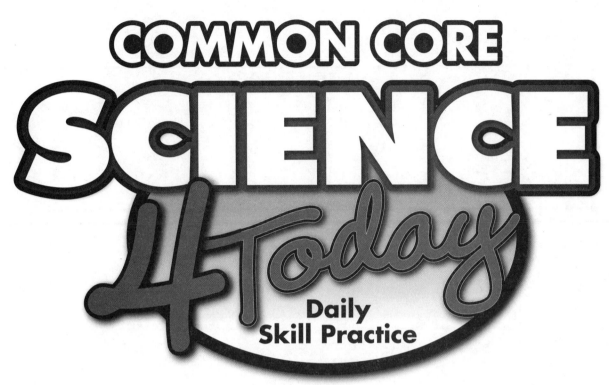

COMMON CORE

SCIENCE

4 Today

Daily Skill Practice

Grade 4

Carson-Dellosa Publishing, LLC
Greensboro, North Carolina

Credits

Content Editor: Elise Craver
Copy Editor: Karen Seberg

 Visit *carsondellosa.com* for correlations to Common Core, state, national, and Canadian provincial standards.

Carson-Dellosa Publishing, LLC
PO Box 35665
Greensboro, NC 27425 USA
carsondellosa.com

ISBN 978-1-4838-1127-7
01-135141151

Table of Contents

Introduction

Common Core Science 4 Today is a perfect supplement to any classroom science curriculum. Students' science skills will grow as they support their knowledge of science topics with a variety of engaging activities.

This book covers 40 weeks of daily practice. You may choose to work on the topics in the order presented or pick the topic that best reinforces your science curriculum for that week. During the course of four days, students take about 10 minutes to complete questions and activities focused on a science topic. On the fifth day, students complete a short assessment on the topic.

Various skills and concepts in math and English language arts are reinforced throughout the book through activities that align to the Common Core State Standards. Due to the nature of the Speaking and Listening standards, classroom time constraints, and the format of the book, students may be asked to record verbal responses. You may wish to have students share their answers as time allows. To view these standards, please see the Common Core State Standards Alignment Matrix on pages 5–8.

Indicates the weekly practice page

Indicates the daily practice exercises

Indicates the weekly assessment

Indicates the Common Core State Standards covered in the daily practice exercises and the weekly assessment

Common Core State Standards Alignment Matrix

English Language Arts

STANDARD	W1	W2	W3	W4	W5	W6	W7	W8	W9	W10	W11	W12	W13	W14	W15	W16	W17	W18	W19	W20
4.RI.1														●						●
4.RI.2																				
4.RI.3				●			●					●					●			
4.RI.4							●					●		●						
4.RI.5																				
4.RI.6																				
4.RI.7						●	●		●				●		●	●				●
4.RI.8																				
4.RI.9																				
4.RI.10														●		●			●	
4.W.1																			●	
4.W.2									●											●
4.W.3												●								
4.W.4																			●	
4.W.5																				
4.W.6																				
4.W.7																				
4.W.8																				
4.W.9																				
4.W.10									●			●							●	
4.SL.1				●	●			●		●	●		●					●		
4.SL.2																				
4.SL.3																				
4.SL.4			●		●												●			
4.SL.5							●							●			●			●
4.SL.6																				
4.L.4	●										●					●		●		
4.L.5					●															
4.L.6	●	●	●	●	●	●		●	●	●	●	●	●		●		●	●	●	●

W = Week

Common Core State Standards Alignment Matrix

English Language Arts

STANDARD	W21	W22	W23	W24	W25	W26	W27	W28	W29	W30	W31	W32	W33	W34	W35	W36	W37	W38	W39	W40
4.RI.1									●						●					
4.RI.2																				
4.RI.3		●		●																
4.RI.4																				
4.RI.5																				
4.RI.6																				
4.RI.7					●	●	●					●				●	●	●		●
4.RI.8																				
4.RI.9																				
4.RI.10													●	●	●					
4.W.1																				●
4.W.2							●			●			●	●	●					
4.W.3	●																			
4.W.4																				
4.W.5																				
4.W.6																				
4.W.7																				
4.W.8																				
4.W.9																				
4.W.10	●																			●
4.SL.1		●			●	●		●								●	●		●	
4.SL.2																				
4.SL.3																				
4.SL.4												●		●				●		
4.SL.5		●																●		
4.SL.6																				
4.L.4		●		●		●		●	●	●		●					●			●
4.L.5																				
4.L.6	●	●	●	●	●	●	●	●	●	●	●	●	●	●	●		●	●	●	●

W = Week

CD-104815 • © Carson-Dellosa

Common Core State Standards Alignment Matrix

Math

STANDARD	W1	W2	W3	W4	W5	W6	W7	W8	W9	W10	W11	W12	W13	W14	W15	W16	W17	W18	W19	W20
4.OA.A.1											●									
4.OA.A.2											●									
4.OA.A.3																				
4.OA.B.4																				
4.OA.C.5																				
4.NBT.A.1															●					
4.NBT.A.2															●					
4.NBT.A.3																●				
4.NBT.B.4																				
4.NBT.B.5															●					
4.NBT.B.6										●						●				
4.NF.A.1																				
4.NF.A.2			●					●												
4.NF.B.3																				
4.NF.B.4										●										
4.NF.C.5																				
4.NF.C.6																				
4.NF.C.7																				
4.MD.A.1	●	●													●	●				
4.MD.A.2		●						●												
4.MD.A.3																				
4.MD.B.4			●														●			
4.MD.C.5																				
4.MD.C.6																				
4.MD.C.7										●										
4.G.A.1																				
4.G.A.2																				
4.G.A.3																				

W = Week

Math

STANDARD	W21	W22	W23	W24	W25	W26	W27	W28	W29	W30	W31	W32	W33	W34	W35	W36	W37	W38	W39	W40
4.OA.A.1																				
4.OA.A.2												●								
4.OA.A.3										●	●					●			●	
4.OA.B.4																				
4.OA.C.5						●														
4.NBT.A.1												●								
4.NBT.A.2																	●			
4.NBT.A.3																				
4.NBT.B.4																●				
4.NBT.B.5					●							●							●	
4.NBT.B.6																				
4.NF.A.1																				
4.NF.A.2																				
4.NF.B.3																				
4.NF.B.4			●		●						●									
4.NF.C.5																				
4.NF.C.6																				
4.NF.C.7						●														
4.MD.A.1										●										
4.MD.A.2										●	●									
4.MD.A.3																			●	
4.MD.B.4																				
4.MD.C.5																				
4.MD.C.6						●														
4.MD.C.7																				
4.G.A.1																				
4.G.A.2																				
4.G.A.3																				

W = Week

CD-104815 • © Carson-Dellosa

Science Tools

Write the name of a science tool to correctly complete each sentence.

1. Tia measures the height of a small houseplant with a _____.
2. Carlos uses a _____ to see if the water is hot enough.
3. Kim finds the mass of a rock using a _____.
4. Lily uses a _____ to measure six milliliters of water.
5. Ty watches the minute and second hands on the _____ to find out how long it takes the ice cube to melt.

Day 1

1. What does a hand lens do?

   _____

   _____

2. Name two ways a scientist might use a hand lens.

   _____

   _____

Day 2

1. Explain how to use a balance to find mass.

   _____

   _____

   _____

   _____

   _____

   _____

Day 3

Tell what each tool does. Give an example of how you would use each one.

1. microscope: _____

   _____

2. telescope: _____

   _____

3. binoculars: _____

   _____

Day 4

Science Tools

Circle the best answer.

1. What does a spring scale measure?
   A. volume
   B. mass
   C. force
   D. length

2. Which units are related to the units on a ruler?
   A. degrees
   B. grams
   C. minutes
   D. kilometers

Answer the questions.

3. Why would a scientist use a camera?

   _____

   _____

4. Why is the computer an important tool? Tell three ways that a scientist could use it.

   _____

   _____

   _____

   _____

   _____

   _____

   _____

   _____

   _____

   _____

4.L.4, 4.L.6, 4.MD.A.1 CD-104815 • © Carson-Dellosa

The Metric System

1. What is the customary system of measurement?

   _____

   _____

2. What is the metric system of measurement?

   _____

   _____

1. What measurement system is used in the science community?

   _____

2. Why do all scientists use this system?

   _____

   _____

   _____

Write the abbreviation for each metric measurement word.

1. liter _____
2. gram _____
3. meter _____
4. centimeter _____
5. milliliter _____
6. kilogram _____
7. millimeter _____
8. kilometer _____

Convert the measurements to make each equation true.

1. 1 L = _____ mL
2. 1 g = _____ cg
3. 1 m = _____ mm
4. 10 mg = _____ g
5. 100 cL = _____ mL
6. 0.1 m = _____ cm
7. 10 cg = _____ mg
8. 10 L = _____ cL

The Metric System

Circle the best answer.

1. About what temperature is a popsicle?
   A. −5°C
   B. 10°C
   C. 20°C
   D. 32°C

2. About how long is a bee?
   A. 2 m
   B. 2 cm
   C. 2 mm
   D. 2 km

Write the name of an object that is about the size of each measurement.

3. 8 centimeters _____

4. 1 meter _____

5. 1 liter _____

6. 100 milliliters _____

7. 1 gram _____

Answer the question.

8. Antone jumped 9 meters, and Wesley jumped 1,000 centimeters. Who jumped farther? Explain.

_____

_____

_____

_____

Name_____

Science Process Skills

Draw a line to match each science process skill with its definition.

1. observing   grouping objects based on characteristics or qualities

2. classifying   using your five senses to learn about the world

3. communicating   telling how objects are alike and different

4. inferring   making an educated guess about what will happen

5. predicting   sharing information using words and visual aids

6. comparing   using what you know and learn to make conclusions

Day 1

You are asked to explain the following science process skills to another student. Write what you will say. Be sure to give an example of how you could use each skill.

1. making and using models: _____

_____

2. making operational definitions: _____

_____

Day 2

1. What is a variable?

_____

_____

2. Why would a scientist need to identify and control variables in an experiment?

_____

_____

_____

Day 3

1. Name two reasons that scientists take measurements.

_____

_____

_____

_____

_____

Day 4

13

Science Process Skills

Answer the questions.

1. Mia was doing an experiment about how much water her dog needs each day. For five days, she gathered data measuring the amount of water her dog drank. The chart below shows her data. Use the information to make a line plot beside the chart.

Day	Amount of Water
Day 1	$\frac{1}{2}$ L
Day 2	$\frac{3}{8}$ L
Day 3	$\frac{3}{4}$ L
Day 4	$\frac{3}{4}$ L
Day 5	$\frac{5}{8}$ L

2. Write two conclusions that Mia can make from the line plot.

    _____

    _____

    _____

    _____

Circle the best answer.

3. Which science process skill did Mia use when she measured the amount of water each day?

    A. observing

    B. predicting

    C. inferring

    D. estimating

4. Which science process skill did Mia use when she made conclusions from the graph?

    A. experimenting

    B. modeling

    C. observing

    D. inferring

4.SL.4, 4.L.6, 4.NF.A.2, 4.MD.B.4

The Scientific Method

1. Write the numbers 1 through 8 to order the steps of the scientific method.

_____ Draw conclusions about the data.

_____ Compare the conclusion with the hypothesis.

_____ Plan the experiment and the variables.

_____ State the problem.

_____ Conduct the experiment.

_____ Collect the data.

_____ Communicate the results.

_____ Make a hypothesis.

1. What is a hypothesis?

_____

_____

2. Circle the best hypothesis.
   A. Does orange juice have more vitamins than apple juice?
   B. Orange juice has more vitamin C than apple juice.
   C. Orange juice tastes better than apple juice.

1. Jay is doing an experiment to find out what kinds of materials will make a good bike reflector. What are three variables he might choose?

_____

_____

2. In the experiment, how many variables will Jay change each time? Explain.

_____

_____

Write **true** or **false**.

1. _____ The problem is always stated as a question.

2. _____ The experiment's conclusion and its hypothesis are always the same.

3. _____ The hypothesis is the final answer in an experiment.

4. _____ Scientists do not like to share the results of their experiments.

5. _____ When scientists interpret the data, they decide what the information means.

The Scientific Method

Answer the questions.

1. Give two reasons why scientists follow the scientific method.

    _____

    _____

    _____

2. Drew's hypothesis and conclusion do not match. Write how you would explain to Drew what steps he should take next and why. Share your explanation with a partner.

    _____

    _____

    _____

3. Why do scientists communicate information about their experiments?

    _____

    _____

    _____

Circle the best answer.

4. Which activity will a scientist not do during an experiment?

    A. follow a set of steps

    B. observe the action

    C. change one variable

    D. explain the action

5. Which step comes after interpreting the data in the scientific method?

    A. drawing conclusions about the data

    B. making a hypothesis

    C. planning the experiment

    D. stating the problem

4.RI.3, 4.SL.1, 4.L.6

Scientific Inquiry

1. Nora believes that the scientific method is the only way to carry out scientific inquiry. Do you agree? Why or why not? Give examples to support your point of view.

_____

_____

_____

_____

1. What are two things scientists do if they do not agree with the results of an experiment?

_____

_____

_____

_____

_____

Write a short definition for each word.

1. hypothesis: _____

_____

2. theory: _____

_____

3. law: _____

_____

1. Give three reasons why it is important for a scientist to carefully record the conditions and steps of an experiment.

_____

_____

_____

_____

_____

Scientific Inquiry

1. Write notes for an oral report to explain how scientific conclusions can change over time as new knowledge is gained. Prepare several examples to support your statement.

_____

_____

_____

_____

_____

_____

_____

_____

_____

_____

_____

_____

_____

_____

_____

_____

_____

_____

_____

_____

_____

_____

4.SL.1, 4.SL.4, 4.L.5, 4.L.6

Name_____

Observation and Inference

1. What is an observation?

_____

_____

2. What is an inference?

_____

_____

1. How are observations and inferences different? How are they similar?

_____

_____

_____

_____

_____

Label each statement **O** for observation or **I** for inference.

1. _____ The mealworms moved toward the light source.

2. _____ Warming eggs with red light helps the chicks inside grow better.

3. _____ Soap bubbles made with glycerin pop an average of four seconds after the normal bubbles.

4. _____ Water from source B is cloudy and has sediment in it.

5. _____ Cold temperatures cause candy with a lot of sugar to shatter more easily.

Write an inference that could be made from each observation.

1. The magnet did not attract the wooden toy.

_____

2. Quartz scratched both talc and chalk.

_____

3. Predatory birds, such as hawks, have sharp, pointed beaks.

_____

Observation and Inference

List one observation and one inference about each picture.

1.

Observation: _____

_____

_____

Inference: _____

_____

_____

2.

Observation: _____

_____

_____

Inference: _____

_____

_____

3.
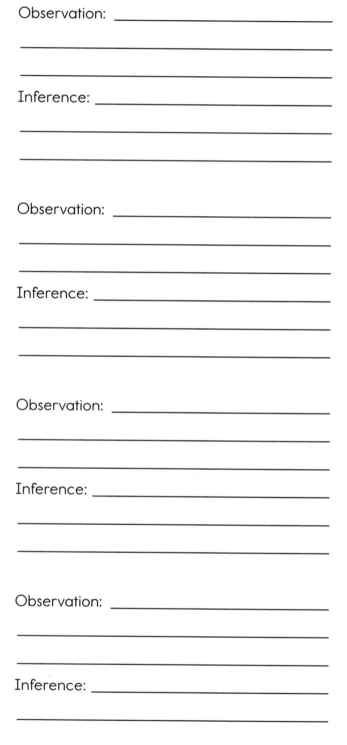

Observation: _____

_____

_____

Inference: _____

_____

_____

4.

Observation: _____

_____

_____

Inference: _____

_____

_____

Name_____

Matter

1. What is matter?

   _____

   _____

2. Think about an orange. Describe four of its physical properties.

   _____

   _____

   _____

1. Which has more mass—a potato or a potato chip? Explain.

   _____

   _____

2. Look at the pictures. Circle the picture that has the greatest volume. Explain your choice.

   _____

   _____

   _____

1. What are the three states of matter? Give an example of each.

   _____

   _____

   _____

2. Look at the pictures in Day 2. What do they show about one property of a liquid?

   _____

   _____

Label each property with **S** for solid, **L** for liquid, or **G** for gas to identify the state of matter it describes. Some properties may describe more than one state of matter.

1. _____ It changes shape easily.

2. _____ Its particles expand to fit the entire container.

3. _____ It does not flow or compress.

4. _____ It takes up space and has mass.

5. _____ Its particles move around but hold on to each other (surface tension).

Matter

Answer the questions.

1. Mason turns a plastic cup upside down and pushes it into a sink filled with water. What happens? Explain.

   _____

   _____

   _____

2. Imagine that you have two different objects that are the same size. Do these items always have the same mass? Explain.

   _____

   _____

   _____

3. You have each state of matter in your body.

   Name a part of your body that is a solid. _____

   Name a part of your body that is a liquid. _____

   Name a part of your body that is a gas. _____

4. Your friend does not understand the differences between the three states of matter. Draw the particles for each state of matter in the boxes below to help you explain the differences. Share your drawings and explanations with a partner.

solid	liquid	gas

Changes in Matter

1. What is a physical change? Give two examples.

   _____

   _____

   _____

2. What is a chemical change? Give two examples.

   _____

   _____

   _____

Write **P** if the example is a physical change. Write **C** if the example is a chemical change.

1. _____ cutting a smaller slice of bread
2. _____ burning wood
3. _____ adding water to soup
4. _____ a car fender rusting
5. _____ ice melting
6. _____ an avocado browning
7. _____ crushing a can
8. _____ stapling paper together
9. _____ dissolving sugar in tea
10. _____ melting silver

1. Cole mixes lemonade powder into a glass of water. His little sister April thinks that the powder disappeared. Tell how Cole can explain to April what happened.

   _____

   _____

   _____

   _____

   _____

   _____

1. Cup A and B have the same amount of water. Cup A has $\frac{2}{3}$ g of powder mixed in. Cup B has $\frac{5}{7}$ g of powder mixed in. Which mixture is more concentrated? How do you know?

   _____

   _____

   _____

Changes in Matter

Answer each question.

1. Liv made three mixtures. Mixture A had $\frac{2}{15}$ L of the substance added, Mixture B had $\frac{5}{6}$ L added, and Mixture C had $\frac{3}{10}$ L added. Write the mixtures in order from most diluted to least diluted. How much more of the substance does the least diluted sample have than the most diluted sample?

2. Hunter does not understand how solutions and mixtures are related. Tell how you would explain the concept to him. Then, share your explanation with a partner.

_____

_____

_____

3. What is the difference between a physical change and a chemical change? Give an example of each.

_____

_____

_____

Circle the best answer.

4. Which is not a physical change?
   A. heating
   B. cooling
   C. writing
   D. tarnishing

5. When does water change from a solid to a liquid?
   A. at its melting point
   B. at its boiling point
   C. at its physical change
   D. at its chemical change

4.SL.1, 4.L.6, 4.NF.A.2, 4.MD.A.2

Energy

1. What is energy?

    _____

    _____

2. Describe three ways you use energy in your everyday life.

    _____

    _____

    _____

Day 1

Draw a line to match each term with its meaning.

1. energy   the energy of an object because it is moving

2. potential energy   the energy caused by a chemical change

3. kinetic energy   the ability to do work

4. mechanical energy   the energy of motion

5. electrical energy   the energy that an object has because of its position

6. chemical energy   the energy caused by the flow of electricity

Day 2

1. Think about a car on a roller coaster. Label the picture to create a diagram for a passage about energy. Write **P** where the greatest potential energy will be for the car. Write **K** where the greatest kinetic energy will be for the car.

Day 3

Write **true** or **false**.

1. _____There are only three types of energy.

2. _____Energy can come in several forms from the same source.

3. _____Eating food gives humans and animals energy.

4. _____Throwing a ball is an example of potential energy.

5. _____Energy is the ability to do work.

6. _____Sound is not a form of energy.

Day 4

Energy

Answer the questions.

1. Create text for a brochure on the types of energy found in a classroom. Name three forms of energy and describe the use of each.

   _____

   _____

   _____

   _____

   _____

2. How are radiant energy waves similar to and different from other kinds of waves?

   _____

   _____

   _____

3. Think about a skier on a snow-covered hill. How does the person change potential energy into kinetic energy?

   _____

   _____

   _____

4. Draw a diagram to demonstrate the potential and kinetic energy of a skier.

Circle the best answer.

5. What kind of energy is an X-ray?
   A. chemical
   B. radiant
   C. mechanical
   D. electrical

4.RI.7, 4.W.2, 4.W.10, 4.L.6

Energy Transformation

1. Can energy be lost or destroyed? Why or why not?

_____

_____

_____

_____

_____

_____

Day 1

Draw a line to match the type of energy transformation to its example.

1. chemical to mechanical   using an electric hot plate

2. electrical to radiant   rubbing hands together to warm them

3. electrical to thermal   using batteries to power a laptop

4. mechanical to thermal   a car burning gasoline to move

5. chemical to electrical   turning on a lamp

Day 2

1. How does energy change when you turn on a lamp?

_____

_____

2. How does energy change when you eat food?

_____

_____

Day 3

1. A pendulum shows the transformation of energy from potential to kinetic and back. Friction from the air slows the pendulum until it eventually stops. In her experiment with a pendulum, Rebecca found that when the pendulum's first swing creates an angle of 100°, it loses 5° with each swing. How many times should the pendulum swing before it stops?

Day 4

Energy Transformation

Answer the questions.

1. Give three examples of energy transformation in your home.

_____

_____

_____

_____

_____

2. Prepare a short statement you would give in a discussion about how energy transformation affects someone's daily life.

_____

_____

_____

_____

_____

3. Kade and his father were hitting baseballs. His father's baseball went 185 feet. Kade's swing had $\frac{3}{5}$ the potential energy of his father's swing. How far did Kade's baseball travel?

Circle the best answer.

4. How does a battery-operated watch change energy?

A. electrical to nuclear

B. mechanical to electrical

C. chemical to mechanical

D. thermal to electrical

Force and Motion

1. What is a force?

   _____

2. Name three ways that a force can affect an object.

   _____

   _____

   _____

   _____

Write the word that completes each sentence.

1. A leaf falls off a tree because of the force of _____.
2. A soccer ball slows down and stops rolling because of the force of _____.
3. A book will stay on a table until it is picked up because of _____.

Answer the question. Then, discuss your answer with a partner.

4. How would you explain to your friend why it is harder to walk up the stairs than to walk down them?

   _____

   _____

Write **true** or **false**.

1. _____ A moving object will always move in a circle.
2. _____ A force will make a moving object slow down or stop.
3. _____ An object that is not moving can sometimes move by itself.
4. _____ All objects have the ability to stay at rest or move.
5. _____ When a force is applied to an object, the object pushes back with the same amount of force.

1. Ella used a spring scale to measure the force of several objects. Object A had a force of 5 newtons (N). Object B had a force 6 times as large as Object A. What was the force of Object B?

Force and Motion

Read each situation. Explain what kind of force is working.

1.   Jayla does a handstand on the gym floor.

   _____

   _____

   _____

2.   Dion catches a soccer ball with his hands before it flies into the net.

   _____

   _____

   _____

3.   The brakes slow Sam's bike.

   _____

   _____

   _____

4.   A roller coaster zooms down a hill.

   _____

   _____

   _____

5.   The bat strikes the ball and sends it flying over the fence.

   _____

   _____

   _____

Circle the best answer.

6.   Which force makes objects slow or stop?
   A. inertia
   B. friction
   C. gravity
   D. magnetic

Electricity

1. Name the three parts of an atom and the kind of charge each part has.

   _____

   _____

2. How does an object get an electric charge?

   _____

   _____

Day 1

Label the diagram.

2. _____

1. _____

4. _____

3. _____

5. What does the diagram show?

   _____

Day 2

1. Explain how the current flows through the system shown in Day 2.

   _____

   _____

   _____

2. If the switch is open, what happens to the system? Explain.

   _____

   _____

Day 3

1. Why are electrical wires covered in plastic? Use the words in the box in your explanation.

conductor	insulator

   _____

   _____

   _____

   _____

Day 4

Electricity

Circle the best answer.

1. What makes an electric charge?

    A. gaining electrons   B. losing electrons

    C. gaining negative charges   D. all of the above

2. Which material is not a conductor?

    A. glass   B. plastic

    C. metal   D. water

Answer the questions.

3. Leo finishes his homework and turns off the lamp. What happens? Explain.

    _____

    _____

4. Ellen is decorating a tree with lights. When she checks the string of lights, she sees that one bulb is burned out, even though the rest of the lights are lit. What kind of circuit does the string have? Explain how you know.

    _____

    _____

    _____

5. How does the use of electricity affect your daily life? Write a short story telling about a day with no electricity. Be sure to include examples and details.

    _____

    _____

    _____

    _____

    _____

    _____

    _____

    _____

    _____

4.RI.3, 4.RI.4, 4.W.3, 4.W.10, 4.L.6   CD-104815 • © Carson-Dellosa

Magnetism

Day 1

1. What is a magnet?

   _____

2. What parts of the magnet have the greatest magnetic force?

   _____

   _____

Day 2

1. The ends of the magnets are called _____.

2. There is a north end and a _____ end.

3. If a magnet is tied to a string and held in the air, the north end of the magnet will point _____.

4. The unlike ends of two magnets _____ each other.

5. The like ends of two magnets _____ each other.

Day 3

1. Look at the picture. Write **N** or **S** to label the ends of the magnets.

2. Explain what is happening in the picture.

   _____

   _____

   _____

   _____

   _____

Day 4

Write **true** or **false**.

1. _____ A compass is a device that has a magnetized needle.

2. _____ The needle in a compass always points south.

3. _____ Earth is a magnet that has a north pole and a south pole.

4. _____ A needle can be magnetized by stroking it along one end of a magnet in the same direction many times.

5. _____ Earth's magnetic north pole is at the North Pole.

Name_____

Magnetism

Answer the questions.

1. List three materials that are not attracted to magnets.

_____ _____ _____

2. How does a compass use a magnet to work?

_____

_____

3. Draw a diagram showing the poles and magnetic field of a bar magnet.

4. Seth is arguing that magnets will not work unless the magnet is touching the metal surface of an item. Is he right or wrong? Write down what you would say to discuss this with him. Be sure to include examples to support your ideas. Discuss your answer with a partner.

_____

_____

_____

_____

_____

_____

_____

_____

_____

_____

Electromagnetism

1. How are electricity and magnetism related?

_____

_____

_____

_____

_____

_____

Unscramble the letters in parentheses to complete each sentence with a device that uses electromagnetism.

1. Some _____ (ordo scolk) need an electrical signal to unlock and open.

2. Some _____(ardh ridesv) in computers use electromagnetism to store and retrieve information.

3. When the power goes out, many homes and businesses use _____(retgneasro) to produce electricity.

4. Special _____(rcnaes) in junkyards can pick up and separate the metal objects from other trash.

1. Draw a diagram of a simple electromagnet made with a battery, wire, and a nail that could accompany a discussion on electromagnets. Be sure to label each part of the electromagnet.

How would you like to travel over 300 miles (482.8 km) per hour—on land? In some countries, people are traveling this fast on maglev trains. While the cars look like those on a train, the maglev train uses electromagnets to move. The guideways that direct the train are lined with metal coils. As electricity moves along the guideways, large magnets on the underside of the car repel the magnetized coil. As a result, the train rises almost four inches (10.16 cm) above the track and glides forward.

1. What poles of the magnets on the train and the magnetized coils are facing each other? Underline the part of the text that supports your answer.

_____

Electromagnetism

Read the paragraph. Then, answer the questions.

In 1820, Hans Oersted was experimenting with electricity. He sent an electric current along a wire. A compass happened to be near the experiment, and Oersted noticed that the needle pointed toward the wire when the current was moving. With more experimenting, Oersted discovered that a magnetic field surrounded any conductor through which a current flowed. Five years later, William Sturgeon found that the force of the magnetic field could be increased if a piece of iron was put inside a coiled wire. It was the first electromagnet. By the end of the 1820s, Joseph Henry made a practical electromagnet. Today, huge electromagnets are built inside generators. They create electricity that powers many appliances in your house. Electromagnets are also inside your home in the form of doorbells and telephones.

1. How does an electromagnet work?

    _____

    _____

    _____

2. Why might an electromagnet be a good source of energy for a doorbell?

    _____

    _____

    _____

3. Why was Hans Oersted a good scientist? Name at least two characteristics.

    _____

    _____

    _____

    _____

4. What parts make up the system of an electromagnet? Underline the part of the passage that helped you answer.

    _____

    _____

    _____

4.RI.1, 4.RI.4, 4.RI.10, 4.SL.5

Name_____

Sound

1. Listen to the sounds around you. Name three sounds that are made by natural objects and three sounds made by human-made objects.

    Natural sounds: _____

    _____

    Human-made sounds: _____

    _____

2. How are all sounds made?

    _____

1. What are sound waves?

    _____

    _____

2. How do sound waves move?

    _____

    _____

    _____

    _____

1. What is volume?

    _____

    _____

2. Why does shouting have more volume than whispering?

    _____

    _____

    _____

1. Circle the sound wave that has a low pitch.

2. Explain why you circled the picture you did.

    _____

    _____

    _____

Sound

Look at the graph. Then, answer the questions.

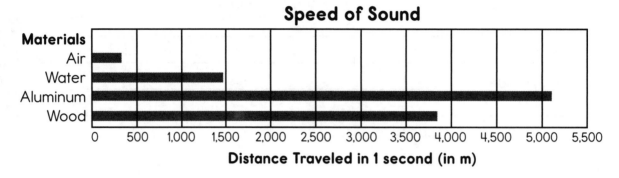

1. Through which states of matter can sound travel?

   _____

2. Of the materials on the graph, through which does sound travel the fastest?

   _____

3. Why might sound travel more quickly through aluminum than air?

   _____

4. If you wanted to use one of the materials above to soundproof a room, which would you choose? Explain.

   _____

   _____

   _____

5. About how far would sound travel in one minute through each material? Complete the chart.

Material	Distance in 1 second* (in m)	Distance in 1 minute (in m)
Air	340	
Water	1,490	
Aluminum	5,100	
Wood	3,850	

   *Distances are rounded to the nearest 10.

Name_____

Light

Day 1

1. Why do scientists refer to **visible light** when they are speaking about light?

_____

_____

_____

_____

_____

_____

Day 2

Write a word from the box to complete each sentence.

| gamma rays   microwaves   radio   ultraviolet   wavelengths   X-rays |

There are other _____ of light that are useful to scientists. Infrared light can be used to see temperature changes. _____ light, also called UV light, is a type of radiation from the sun. It can give you skin cancer, so it is important to use sunscreen that blocks it. We use _____ to cook food. _____ waves carry sound information and are also given off by stars. Doctors and dentists use _____ and _____ to help see inside your body. Even though they are not visible, the different types of waves on the electromagnetic spectrum are used in many ways.

Day 3

1. Write a caption to support this diagram.

| Gamma Rays | X-Rays | Ultraviolet Rays | Infrared Rays | Microwaves Radar | Radio |

Visible Light

| Violet | Blue | Green | Yellow | Orange | Red |

_____

_____

Day 4

The speed of light in a vacuum is 299,792,458 meters per second, or about _____ kilometers per second. Scientists use the speed of light to measure large distances in light years. A **light year** is 5,865,696,000,000 miles (9,460,800,000,000 km).

A **light nanosecond** is the distance light can travel in a billionth of a second and can measure the distance of closer objects. A light nanosecond is about 1 foot (30 cm). Radar sends out a radio pulse that bounces off of the object and returns. Since radio waves travel at the speed of light, scientists can divide the number of nanoseconds by 2 to find an object's distance.

1. If a radio pulse takes 8,948 nanoseconds to return from an airplane, how many feet away is the airplane? _____

Light

Answer the questions.

1. How does light travel?

   _____

   _____

2. Draw a diagram to show how light behaves in each situation.

reflected	refracted	absorbed

3. How are color and light related?

   _____

   _____

   _____

Circle the best answer.

4. Light energy can be converted into _____.

   A. heat

   B. motion

   C. chemical

   D. all of the above

5. Light and sound energy both _____.

   A. travel in waves

   B. can come from the sun

   C. rely on vibrations

   D. are potential energy

Name_____

Plants

1. Write a **P** in front of each characteristic of a plant.

    _____ can grow   _____ learns behavior

    _____ made of cells   _____ can reproduce

    _____ needs shelter   _____ gets foods from dead organisms

    _____ makes its own food   _____ cannot move

1. What are the two main groups of plants? Give two examples of each.

    Group 1: _____

    _____

    Group 2: _____

    _____

2. Does a daisy belong in the same main plant group as a pine tree? Explain.

    _____

    _____

Prepare for a presentation on plants. Label the parts of a flower. Tell what each part does. Discuss with a partner.

3. _____

1. _____

4. _____

2. _____

1. How does moss reproduce?

    _____

    _____

2. What is pollination? Name two ways that plants are pollinated.

    _____

    _____

    _____

    _____

Plants

Answer the questions.

1. Explain fertilization of a flowering plant. Use the words in the box in your explanation.

| egg | ovary | pistil | pollen | seed | sperm |

_____

_____

_____

_____

2. How do insects pollinate flowers?

_____

_____

3. A corn plant grows two kinds of flowers. The flowers that have the stamens grow at the top of the stalk. The flowers that have the pistils grow on the rest of the stalk. How does the plant reproduce?

_____

_____

4. Carter did an experiment on how quickly bean plants grow. He measured 12 plants after a month to the nearest $\frac{1}{2}$ centimeter. Use the data in the table to create a line plot.

11	13	$12\frac{1}{2}$	$11\frac{1}{2}$
$12\frac{1}{2}$	$9\frac{1}{2}$	11	$12\frac{1}{2}$
13	12	$12\frac{1}{2}$	$13\frac{1}{2}$

Circle the best answer.

5. What is the name of the process where seeds begin to grow?

   A. fertilization

   B. germination

   C. pollination

   D. classification

Name_____

Animals

Unscramble the letters in parentheses to make words that tell about animal characteristics.

1. Animals can grow and _____(druepcroe).

2. They all must find their own _____(dofo).

3. Some live in a _____(reelths) that protects them from weather and predators.

4. Adult animals have _____(noyug) that look like them.

5. All animals need _____(nyxoge) to breathe.

Day 1

1. What are the two main groups of animals? Give two examples of each.

   Group 1: _____

   _____

   Group 2: _____

   _____

2. How are a beetle and worm alike? How are they different?

   _____

   _____

Day 2

Animals with backbones can be classified into smaller groups, or classes. Draw a line to match each class with one of its defining characteristics.

1. fish   feeds its young milk

2. amphibian   has a body covered in dry scales

3. reptile   lives in freshwater or salt water

4. bird   has a body covered in feathers

5. mammal   lives part of its life in the water and part of it on land

Day 3

1. What is an exoskeleton? Explain its function.

   _____

   _____

2. What does it mean to be cold-blooded? Name three classes of animals that share this characteristic.

   _____

   _____

   _____

Day 4

Name_____

Animals

Answer the questions.

1. What is a backbone? Explain why it is an important characteristic to scientists who study animals.

   _____

   _____

   _____

2. Read the animal names in the box. Write each name in the correct class below.

alligator	bass	frog	hawk	human
parrot	salamander	shark	snake	whale

   Fish   Reptile   Amphibian

   _____   _____   _____

   _____   _____   _____

   Bird   Mammal

   _____   _____

   _____   _____

3. Leigh thinks that because butterflies and hummingbirds both have wings and come from eggs, they are both birds. Tell how you would explain why they do not belong to the same class of animals. Discuss your explanation with a partner.

   _____

   _____

   _____

   _____

Circle the best answer.

4. Which is not a characteristic of a mammal?

   A. Hair can be found on its body.   B. It gives birth to live young.

   C. The female feeds milk to its young.   D. It is cold-blooded.

Name_____

Heredity and Diversity

Write a word from the box to complete each definition.

| diversity | heredity | inherit | species | trait |

1. _____ – the passing of characteristics from one generation to the next

2. _____ – the condition of being different

3. _____ – a group of animals that can reproduce

4. _____ – a feature or characteristic gotten from a parent

5. _____ – to get a characteristic from a parent or ancestor

Puppies look similar to an adult dog when they are young. Yet, puppies in the same litter can look different from one another. They may be different colors. Some may have smooth fur, while others may have rough fur.

1. Think of other species of animals. What are two advantages of animals in a species being different?

_____

_____

2. What are two disadvantages of animals in a species being different?

_____

1. Why is it important that animals inherit traits from their parents?

_____

_____

2. What are three specific traits that you inherited from your parents that your friends did not inherit from their parents?

_____

_____

_____

A species can differ in several ways. Some animals have inherited a characteristic, but individual conditions affect each animal's growth and development. For example, all of the people in a family may be tall. The children in the family will probably be tall too. A person's body systems and the environment will affect how tall that person grows, though.

1. Think about other animals. What are two inherited characteristics that can be affected by other conditions? Explain.

_____

_____

Heredity and Diversity

Answer the questions.

1. How does heredity affect a species?

   _____

   _____

2. What are three main ways animals within a species differ? Give an example of each.

   _____

   _____

   _____

   _____

   _____

3. Do you think it is better to have diversity in a species or not? Write a letter to a science magazine explaining your position. Be sure to state your opinion clearly and support it with specific reasons and facts.

   _____

   _____

   _____

   _____

   _____

   _____

   _____

   _____

   _____

   _____

   _____

   _____

   _____

   _____

4.RI.10, 4.W.1, 4.W.4, 4.W.10, 4.L.6

Name_____

Adaptation

1. What is an adaptation?

    _____

    _____

2. Choose an animal. What is one way its body is adapted to its environment? What is one behavior that helps it adapt to its environment?

    _____

    _____

Write **B** if the adaptation is a behavior. Write **S** if the adaptation is related to structure.

1. _____ thick fur

2. _____ webbed feet

3. _____ flies away for winter

4. _____ builds a nest

5. _____ hisses at unknown animals

6. _____ uses sunlight for energy

1. Juan went to the zoo. He saw an animal with big ears and short brown fur that was peeking out of a hole in the ground. What are four inferences that Juan can make about the adaptations of the animal?

    _____

    _____

    _____

    _____

    _____

1. What is a mutation?

    _____

    _____

2. How can a mutation positively affect a species? How can it have a negative effect?

    _____

    _____

    _____

Adaptation

Answer the questions.

1. Draw a picture to support a report about camouflage. Write a caption for the picture.

    _____

    _____

2. A brown bear and a squirrel belong to different species. Yet, they have several adaptations that are the same. Identify one structural and one behavioral adaptation showing how they are alike. Then, explain how each adaptation helps them.

    _____

    _____

    _____

3. What is one inference that you can make based on the adaptations you named in Question 2?

    _____

    _____

4. Look at the illustration. What are two adaptations that help this organism live in its environment? Explain.

4.RI.1, 4.RI.7, 4.W.2, 4.SL.5, 4.L.6

Name_____._____

Animal Behaviors

1. What is a behavior?

   _____

2. Write a behavior each organism exhibits.

   dog _____

   fish _____

   ant _____

   human _____

Write **L** if the behavior is learned. Write **I** if the behavior is an instinct.

1. _____ A dolphin jumps through a hoop.

2. _____ A butterfly gets nectar from a flower.

3. _____ A fawn stays in the grass until called by her mother.

4. _____ A bird makes a nest.

5. _____ A lion attacks the neck of an antelope.

6. _____ A dog comes when he hears his name.

1. What is a reflex? Give one example.

   _____

   _____

2. How are a stimulus and a response related? Give an example.

   _____

   _____

   _____

1. What are two ways that animals learn behaviors? Give an example of each.

   _____

   _____

   _____

   _____

   _____

   _____

Animal Behaviors

Circle the best answer.

1. Which is a behavior that is an instinct for a baby?

    A. crying

    B. laughing

    C. talking

    D. chewing

Answer the questions.

2. How do instincts help animals? Give two examples.

    _____

    _____

    _____

3. Pretend you are a lion. Write a paragraph from the point of view of a lion, explaining how you use your instincts and learned behaviors to get food.

    _____

    _____

    _____

    _____

    _____

    _____

    _____

    _____

    _____

    _____

    _____

    _____

    _____

    _____

Organisms and Energy

1. What is the process by which plants get food energy? Identify the three things plants need in the process.

   _____

2. How does a plant use its leaves in the process?

   _____

   _____

   _____

Unscramble the letters in parentheses to complete each sentence.

1. All living things need _____(greeny) to survive.
2. Plants use light energy to make _____(guras) that they store.
3. An _____(evibrroeh) is an animal that eats plants for energy.
4. A _____(anorvecri) is an animal that eats other animals for energy.
5. An _____(ermovnio) is an animal that gets energy from both plants and animals.

1. What are two ways that the energy stored in dead organisms is used? Give examples of each.

   _____

   _____

   _____

2. What is the relationship between decomposers and plants?

   _____

   _____

1. Write 1 through 6 to show how energy is passed along in a food chain.

   _____ spider

   _____ cricket

   _____ sun

   _____ snake

   _____ grass

   _____ bird

Organisms and Energy

Answer the questions.

1. Draw an example of a food chain with at least four organisms to support a discussion about food chains. Be sure to label the picture.

2. Explain how human beings can affect the food chain you created.

   _____

   _____

   _____

   _____

3. Write a short definition for each word or phrase.

   decomposer _____

   photosynthesis _____

   food web _____

   consumer _____

   producer _____

Circle the best answer.

4. What is the main source of energy for plants?

   A. sugar   B. nutrients

   C. decomposers   D. oxygen

5. How is energy passed from one organism to another?

   A. chlorophyll   B. food chains

   C. photosynthesis   D. oxygen

Ecosystems

1. What is an ecosystem?

   _____

2. List three ecosystems.

   _____

3. Choose one ecosystem you listed above and name three living and three nonliving things in it.

   Living: _____

   Nonliving: _____

1. Look at the living and nonliving things you listed in Day 1. Explain how they are all a useful part of an ecosystem.

   _____

   _____

   _____

   _____

   _____

   _____

1. What are three changes, natural or human made, that might happen in an ecosystem?

   _____

   _____

   _____

2. How do the above changes affect an ecosystem?

   _____

   _____

1. What is a habitat?

   _____

   _____

2. How is a habitat different from an ecosystem?

   _____

   _____

   _____

Ecosystems

Answer the questions.

1. In some towns, foxes are eating out of garbage cans instead of finding food in forests. Explain why.

   _____

   _____

2. Describe how both living and nonliving things are important parts of a desert ecosystem.

   _____

   _____

   _____

3. Grace is looking at the effect of humans on a nearby ecosystem. Local scientists say that the new shopping centers that were built decreased the population of each species by $\frac{2}{3}$ the next year. If this trend continues, how many of each species will be left in the ecosystem in 2015?

Animal	Population in 2014	Population in 2015
squirrels	1,536	
cardianls	642	
robins	903	
rabbits	594	
chipmunks	708	

4. A town wants to build a grocery store where a nearby woodland is. How would this affect the local ecosystem? You are going to speak to the town council about the effect of building the grocery store. Make notes below about what information you will cover. Be sure to include specific information and examples to support your points.

   _____

   _____

   _____

   _____

   _____

   _____

   _____

Name_____

Cycles in Organisms

1. What is a cycle?

_____

_____

Unscramble the letters in parentheses to make a word that names a cycle.

2. A _____ (feil) cycle describes how an organism grows and reproduces.

3. The _____ (yoxeng) cycle explains how producers and consumers make the air they need to survive.

4. Plants use _____ (stesithhoopnys) to make sugar for energy.

1. What cycle is shown? Write **1** through **5** to show the order.

Cycle name: _____

_____   _____   _____   _____   _____

1. Describe the life cycle of the monarch butterfly. Use the words in the box in your paragraph.

| adult | caterpillar | chrysalis | egg | larva | metamorphosis | pupa |

_____

_____

_____

_____

1. Think about the cycle of the seasons and the life cycle of organisms. How are they related to each other? How are they independent of each other? Give an example for each.

_____

_____

_____

_____

Name_____

Cycles in Organisms

Circle the best answer.

1. Ellen sees a cocoon. What part of the butterfly life cycle does she see?
   A. egg
   B. pupa
   C. caterpillar
   D. butterfly

2. Why do organisms have life cycles?
   A. to reproduce
   B. to eat
   C. to grow
   D. to sense

Answer the questions.

3. Describe the life cycle of a frog.

   _____

   _____

   _____

   _____

   _____

4. Food chains and food webs show a type of cycle. Describe how they are cycles and how they are important to organisms.

   _____

   _____

   _____

   _____

   _____

   _____

Changing Populations

Draw a line to match each word to its meaning.

1. ecosystem   the place an animal lives where all of its needs can be met

2. habitat   all of the populations that live in a place

3. population   all of the living and nonliving things in a place

4. environment   a group of one kind of living thing that lives in a place

5. community   everything that is around a living thing

1. Why do scientists keep track of populations? Explain.

_____

_____

2. Animals often move around. How might scientists keep track of the animals in these populations?

_____

_____

A scientist made a graph to show the changing populations of rabbits and coyotes in a park. Look at the graph. Then, answer the questions.

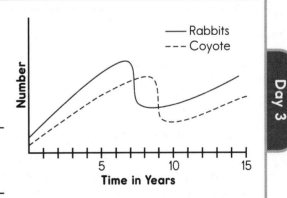

1. What happened to the rabbit population for the first six years?

_____

2. When did the coyote population begin to increase?

_____

1. Look at the graph in Day 3. What can you conclude about the relationship of the rabbit and coyote populations? Explain.

_____

_____

_____

_____

Name_____

Changing Populations

Answer the questions.

1. Your neighbor is surprised that there are fewer birds visiting her bird feeder than there used to be. Write a short response that explains three reasons that a population might change. Discuss your response with a partner.

_____

_____

_____

_____

_____

2. Look at the food chain below. Suppose a disease strikes the mice. Explain what might happen to each remaining organism in the food chain.

_____

_____

_____

_____

3. Local scientists studied the changing populations of owls and mice in a nearby forest. For the first three years, the population of each species doubled. Then, for the next two years, each population decreased by $\frac{1}{8}$. Use the information to complete the population chart.

Year	0	1	2	3	4	5
Owls	208	416				
Mice	532	1,064				

Circle the best answer.

4 Which is a factor that could affect a population?

   A. reproduction rate   B. climate

   C. disease   D. all of the above

 4.RI.7, 4.SL.1, 4.L.6, 4.NBT.B.5, 4.NF.B.4   CD-104815 • © Carson-Dellosa

Earth and the Sun

Write a word to complete each sentence.

1. Earth makes a complete _____ around the sun once each year.

2. Each day, Earth completes one _____.

3. Earth is the _____ planet from the sun.

4. The sun is a ball of burning gases, called a _____.

5. Earth is tilted on its _____, which affects the seasons.

1. Label the diagram.

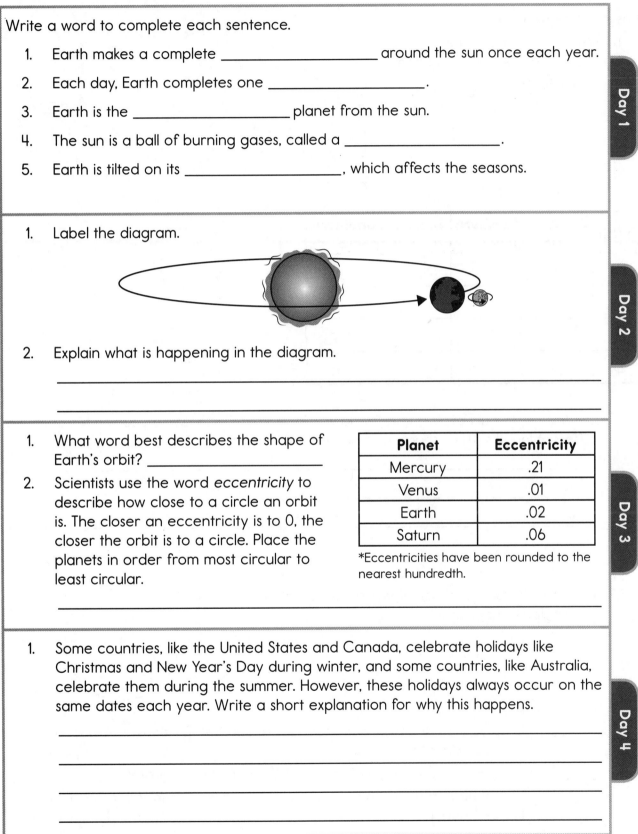

2. Explain what is happening in the diagram.

_____

_____

1. What word best describes the shape of Earth's orbit? _____

2. Scientists use the word *eccentricity* to describe how close to a circle an orbit is. The closer an eccentricity is to 0, the closer the orbit is to a circle. Place the planets in order from most circular to least circular.

Planet	Eccentricity
Mercury	.21
Venus	.01
Earth	.02
Saturn	.06

*Eccentricities have been rounded to the nearest hundredth.

_____

1. Some countries, like the United States and Canada, celebrate holidays like Christmas and New Year's Day during winter, and some countries, like Australia, celebrate them during the summer. However, these holidays always occur on the same dates each year. Write a short explanation for why this happens.

_____

_____

_____

_____

Earth and the Sun

Answer the questions.

1. Write the season on each hemisphere for each diagram of Earth.

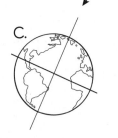
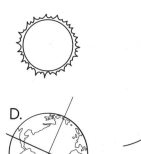
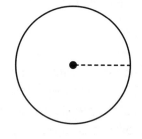

Earth	Season in Northern Hemisphere	Season in Southern Hemisphere
A.		
B.		
C.		
D.		

2. Tell how you would explain to a friend why a year on Earth and a year on Mars are different lengths. Then, share your answer with a partner.

_____

_____

_____

_____

3. Earth's tilt is about 24°. Use a protractor to sketch the angle on the planet.

Write **true** or **false**.

4. _____ Leap year must occur every four years because Earth takes a little longer than 365 days to orbit the sun.

5. _____ The imaginary line Earth spins around is called the *equator*.

6. _____ When the northern hemisphere is tilted toward the sun, it is summer in North America.

The Moon

Write **true** or **false**. Rewrite any false statement to make it true.

1. _____ The moon is a satellite that revolves around Earth.

2. _____ It takes exactly 29 days for the moon to circle around Earth.

3. _____ The moon makes its own light.

4. _____ The changes in the moon's shape as seen from Earth are called phases.

Draw a line to match each phase with its description. Then, write numbers 1 through 8 to show the correct order of the moon's phases.

1. _____ waxing gibbous moon   The moon looks like a big, bright circle.
2. _____ third-quarter moon   A sliver of the shrinking moon is lit.
3. _____ waning gibbous moon   A sliver of the growing moon is lit.
4. _____ waxing crescent moon   One-half of the shrinking moon is lit.
5. _____ first-quarter moon   The moon looks dark in the night sky.
6. _____ waning crescent moon   The growing moon's surface is mostly lit.
7. _____ new moon   One-half of the growing moon is lit.
8. _____ full moon   The shrinking moon's surface is mostly lit.

1. Danny looks out at the night sky. He sees a waxing gibbous moon. Draw a circle in the diagram below to show where the moon is in relation to Earth and the sun.

1. What kind of eclipse is shown in the diagram below?

   _____

2. Explain what is happening in this kind of eclipse.

   _____

   _____

   _____

   _____

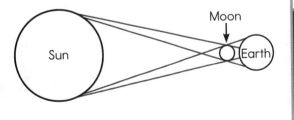

Moon

Sun   Earth

The Moon

Answer the questions.

1. What is the difference between a solar and a lunar eclipse?

   _____

   _____

   _____

2. What is a new moon? Write a short report to explain what it looks like and why it has this appearance.

   _____

   _____

   _____

   _____

   _____

3. The lunar cycle lasts about _____ days. If the first full moon of the year is on January 18 of a nonleap year, when will the next 11 full moons fall? Use a pattern to complete the chart.

Full Moon	Date	Full Moon	Date
1	January 18	7	
2		8	
3		9	
4		10	
5		11	
6		12	

4. Darius sees the moon shown below. What is the name of the moon he sees? Which phase of the moon will he see several nights later?

   _____

4.RI.7, 4.W.2, 4.L.6, 4.OA.C.5 CD-104815 • © Carson-Dellosa

Rocks and Minerals

1. What is a mineral?

    _____

2. Write **P** in front of the properties that help scientists classify minerals.

    _____ how hard it is   _____ its color

    _____ how shiny it is   _____ its temperature

    _____ how it breaks   _____ if it is magnetic

    _____ how it smells   _____ its size

What are the three kinds of rocks? Tell how each is formed.

1. _____ are formed _____

    _____

2. _____ are formed _____

    _____

3. _____ are formed _____

    _____

Circle the word that best completes each sentence.

1. A (fossil, rock) is the remains or mark left behind by a living thing that died long ago.

2. These kinds of remains are often found in (metamorphic, sedimentary) rocks.

3. The (imprint, cast) shows the mold of something thin, like a wing or leaf.

4. Some fuels we use today, like (ore, coal), are actually fossils formed from plants that lived long ago.

5. The cooled magma from volcanoes forms (igneous, metamorphic) rocks.

1. What is soil made of?

    _____

    _____

2. How is soil made?

    _____

    _____

    _____

Rocks and Minerals

Answer the questions.

1. Explain the ways that rocks can be changed throughout the rock cycle.

     _____

     _____

     _____

     _____

     _____

     _____

2. What facts and details would you use to support a discussion on why soil is important? Compare your answer with a partner's.

     _____

     _____

     _____

     _____

     _____

     _____

Circle the best answer.

3. Which kind of soil is made by decomposers?

     A. clay

     B. minerals

     C. humus

     D. silt

4. Thad picks up a rock that easily breaks in his hands. What kind of rock is it?

     A. sedimentary

     B. volcanic

     C. metamorphic

     D. igneous

Name_____

Fossils

Write a word to complete each sentence telling about fossils.

1. A _____ is a hollow shape that remains where the organism was.

2. A _____ is an animal that lived millions of years ago but is now extinct.

3. Some organisms were trapped in the resin of a tree that then hardened into _____.

4. People who study fossils are _____.

5. A _____ is the shape of an organism that is filled with sediment.

1. Write numbers **1** through **5** to show how a fossil is made.

   _____ The soft parts rot.

   _____ Layers of small rocks, sand, and mud cover the organism.

   _____ The plant or animal dies.

   _____ A print of the organism remains in the rock.

   _____ The pressure of the layers of sediment forms rock.

1. Identify two reasons that scientists study fossils.

   _____

   _____

2. How can knowing about organisms living today help scientists understand about life long ago?

   _____

   _____

   _____

1. What kind of rock is shown here?

   _____

2. Explain how you know.

   _____

   _____

   _____

   _____

Fossils

Answer the questions.

1. How can studying fossils help scientists learn about Earth's history? Give three examples.

    _____

    _____

    _____

    _____

    _____

2. Order the geological time from oldest to the most recent.

    _____ Cenozoic

    _____ Precambrian

    _____ Paleozoic

    _____ Mesozoic

3. A scientist finds the fossil of a fern in Antarctica. What can she infer from this discovery? Give an example to support your inference.

    _____

    _____

    _____

    _____

    _____

Circle the best answer.

4. What is coal made from?
    A. ancient ferns
    B. tree resin
    C. tar
    D. animal bones

Weathering and Erosion

Day 1

Tell how each event changes the surface of Earth.

1. volcano: _____

_____

2. earthquake: _____

_____

3. flood: _____

_____

Day 2

Write **C** if the statement describes chemical weathering. Write **P** if the statement describes physical weathering.

1. _____ A cave forms under the ground in limestone rock.
2. _____ Plant roots grow in the crack of a rock and force the rock to crack.
3. _____ Air pollution wears away the nose of a statue.
4. _____ Water trickles in the crack of a rock and freezes, making the crack bigger.
5. _____ Wind picks up particles of sand and blasts the rock, forming an arch.
6. _____ The acid in a moss begins to wear a hole in a rock.

Day 3

1. What is the difference between weathering and erosion?

_____

_____

2. What are the three forces that cause erosion?

_____

3. How does a glacier cause erosion?

_____

_____

Day 4

Circle the word that best completes each sentence.

1. (Weathering, Deposition) is the settling of rocks and soil once erosion has taken place.
2. The eroded materials fall out once the water or wind (quickens, slows).
3. The soil in these areas is filled with rich (nutrients, gases) that are good for crops.
4. Depositions caused by wind can form (mountains, dunes).
5. Depositions caused by ocean water can form new (bays, beaches).
6. In a river, deposition can create a (delta, canyon).

Weathering and Erosion

Answer the questions.

1. What are four ways that rocks are physically weathered to form soil? Write a short report to explain each way.

    _____

    _____

    _____

    _____

    _____

2. Niagara Falls is a large waterfall on the border of the United States and Canada. Scientists predict that one section of the falls will erode 3 to 4 inches every 10 years. About how long will it take for the falls to erode 2 feet?

3. How can a piece of rock from a mountain end up in a river delta? Use the words in the box in your explanation.

deposition	erosion	weathering

    _____

    _____

    _____

    _____

Circle the best answer.

4. Which helps prevent soil erosion?

    A. spraying water

    B. plowing fields

    C. planting grass

    D. constructing buildings

4.W.2, 4.L.4, 4.L.6, 4.OA.A.3, 4.MD.A.1, 4.MD.A.2

Natural Disasters

Write a word that correctly completes each sentence.

1. A _____ is a mountain from which melted rock flows.

2. An _____ is when the pressure of heat and gas forces magma to the surface.

3. The melted rock that flows out is called _____.

4. Huge columns of clouds made of rocks, gas, and _____ rise high into the air.

5. Many volcanoes are _____, or sleeping, but may erupt someday.

Day 1

1. What causes an earthquake?

   _____

   _____

2. How does an earthquake change the land? Give two examples.

   _____

   _____

   _____

   _____

Day 2

1. How do floods harm the land?

   _____

   _____

2. How can some floods help the land?

   _____

   _____

   _____

   _____

Day 3

1. What are the similarities and differences between a tornado and a hurricane?

   _____

   _____

   _____

   _____

   _____

   _____

Day 4

Natural Disasters

Answer the questions.

1. Will Earth look the same 1,000 years from now? Explain.

_____

_____

_____

2. Which is easier to predict—a tornado or a hurricane? Explain.

_____

_____

_____

3. Volcanoes formed the Hawaiian Islands long ago. In fact, lava still flows out of some volcanoes today. Scientists also watch underwater volcanoes that are erupting. What do you predict will happen to Hawaii thousands of years from now?

_____

_____

Scientists use the Fujita scale to categorize the severity of tornadoes and the Saffir-Simpson Hurricane Wind Scale to categorize the severity of hurricanes. Use the scales to answer the questions.

Enhanced Fujita Scale	
Fujita Number	**Wind Gust Speed (mph)**
F0	65–85 (105–137 km/h)
F1	86–110 (138–177 km/h)
F2	111–135 (179–217 km/h)
F3	136–165 (219–266 km/h)
F4	166–200 (267–322 km/h)
F5	over 200 (322 km/h)

Saffir-Simpson Hurricane Wind Scale	
Category	**Sustained Wind Speed (mph)**
1	74–95 (119–153 km/h)
2	96–110 (154–177 km/h)
3	111–129 (178–208 km/h)
4	130–156 (209–251 km/h)
5	over 156 (251 km/h)

4. Which category of hurricane has wind speeds that are about twice as fast as an F0 tornado? _____

5. When a hurricane with wind speeds of about 150 mph (241 km/h) hit land, its wind speed decreased by $\frac{1}{5}$. What category was it demoted to? _____

6. A tornado with wind gusts of 128 mph increased in power by $\frac{1}{5}$ every 5 minutes. How long will it take the tornado to reach F5 status? _____

Name_____

Weather

Day 1

1. Why does air temperature rise?

   _____

   _____

2. Why is the temperature over land warmer than over water?

   _____

   _____

   _____

Day 2

1. Explain how wind forms.

   _____

   _____

2. What is humidity?

   _____

   _____

   _____

Day 3

1. How do clouds form?

   _____

   _____

2. In May, Kim sees some big, puffy clouds high above in the blue sky. Explain to her why it is or is not a good day for her to have a picnic with her friends.

   _____

   _____

Day 4

1. Look at the picture. What does it show? Describe what is happening.

   warm air   cold air

   _____

   _____

   _____

   _____

   _____

   _____

Weather

Answer the questions.

1. Look at the weather map. Imagine that you are a weather reporter on the news. Tell what you will say in a report.

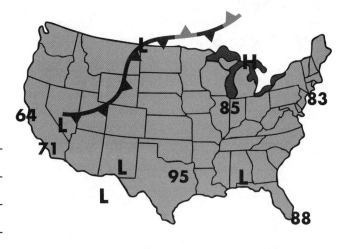

_____

_____

_____

_____

_____

_____

_____

_____

2. Use the information given to complete the chart.

Type of Cloud	Typical Altitude
stratus	0–_____ ft.
stratocumulus	2,000–_____ ft.
cumulus	2,000–_____ ft.
cumulonimbus	2,000–_____ ft.
altocumulus	_____–_____ ft.
altostratus	6,500–_____ ft.
nimbostratus	2,000–18,000 ft.
cirrocumulus	_____–45,000 ft.
cirrus	16,500–_____ ft.
cirrostratus	20,000–_____ ft.

- Cumulus clouds occur within a 1,000-foot range.
- Stratocumulus and stratus clouds end where altocumulus clouds begin.
- Altocumulus clouds end 9 times higher than cumulus clouds begin.
- Cirrostratus clouds end 7 times higher than nimbostratus clouds begin.
- Cumulonimbus and cirrus clouds end 50 times higher than 900 feet.
- Altocumulus clouds begin 500 feet higher than 3 times the lower limit of cumulus clouds.
- Altostratus clouds occur in a 10,000-foot range.
- Cirrocumulus clouds begin where altostratus clouds end.

Name_____

Cycles on Earth

Day 1

1. What is a cycle?

   _____

   _____

2. Write **C** beside each cycle in nature.

   _____ a volcano erupting   _____ a flowing river   _____ petroleum

   _____ a day   _____ the rain   _____ tides

   _____ the seasons   _____ the soil

Day 2

1. Use the words from the box to complete the paragraph.

   | climate   erode   evaporates   life   rock   sediment   water   webs |

   The ocean plays a major role in many of Earth's cycles. It contains most of the
   _____ on the planet. The water vapor that _____ from the ocean
   creates clouds as part of the water cycle. The ocean is also a biome that supports
   many animals and their _____ cycles and food _____. The ocean helps
   to regulate Earth's _____ by absorbing and releasing heat around the planet.
   On the shorelines, waves _____ the land and create _____ as part of
   the _____ cycle. Many of Earth's cycles depend on the ocean.

Day 3

1. Explain the water cycle.

   _____

   _____

   _____

   _____

2. Why is the water cycle important?

   _____

   _____

Day 4

Write **true** or **false**.

1. _____ The carbon cycle follows the exchange of only carbon dioxide in the
   atmosphere.

2. _____ Volcanoes are often a part of the rock cycle.

3. _____ Weather systems do not follow a cycle.

4. _____ Cycles may change slowly over time.

5. _____ Energy from the sun is needed to make the water cycle work.

6. _____ Not all animals have a life cycle.

Cycles on Earth

Circle the best answer.

1. Which step is not a part of the rock cycle?
   A. weathering
   B. compaction
   C. fire
   D. erosion

2. Mark looks at the moon and sees this shape. What phase of the cycle does he see?

   A. waxing gibbous
   B. new moon
   C. three-quarter moon
   D. waning crescent

Answer the question.

3. Choose a cycle that has been identified this week. Write a report that describes it and explains its importance in nature.

_____

_____

_____

_____

_____

_____

_____

_____

_____

_____

4.RI.10, 4.W.2, 4.L.4, 4.L.6

Name_____

Information Technology

1. What is technology?

   _____

   _____

2. List four ways you use technology each day.

   _____

   _____

   _____

1. Write a paragraph explaining how people use a computer to communicate. Give at least three examples to support your explanation.

   _____

   _____

   _____

   _____

   _____

   _____

With a cell phone, people can make phone calls and receive phone calls wherever they are—at the store, at the park, in the car, and so on. Cell phones turn the human voice into a kind of electricity. The electricity travels through the air to a cell phone tower. The tower locates the other phone and sends the electricity there. Once the electricity enters the other phone, the phone changes it back into the human voice.

1. What kind of phone did your parents use when they were young? How has the phone changed?

   _____

   _____

If you turn on the television, you might be able to choose from over 100 programs. There are two main ways that people are able to view so many programs. One way is through cable television. Wires that carry television signals run to each house. People hook their televisions to the wires. Some people may use a satellite dish to receive programs. A dish is a bowl-shaped device that attaches to the outside of a house. A satellite in space collects signals from a television station, makes them louder, and sends them back to Earth instantly.

1. Why might all houses not have cable television? Name two reasons.

   _____

Information Technology

Answer the question.

1. How has technology changed communication in the last 100 years? Write notes for an oral report about technology and communication. Be sure to include examples and details to support your points.

_____

_____

_____

_____

_____

_____

_____

_____

_____

_____

_____

_____

_____

_____

_____

_____

_____

_____

_____

_____

_____

Name_____

Engineering

Engineers use research, science, technology, math, and creativity to create solutions to everyday problems. They design and build things that can be useful in many areas of life. Like scientists, engineers follow a process. They start with a problem they would like to solve. Then, they research, brainstorm, plan, and create a sample. They test and improve their design. Sometimes, they start over! Engineers help improve our daily lives.

1. How are engineers and scientists the same? How are they different? Underline the text in the passage that helped you answer. _____

_____

Day 1

NASA scientists look at specific problems and find ways to solve them. They develop materials and tools that make space travel safer and easier. Freeze-dried food is a NASA invention. Food that is freeze-dried has no moisture. As a result, it is very lightweight, stores compactly, and does not need refrigeration. Moreover, freeze-dried food can last for years. Today, these packets of food are popular with hikers and campers.

1. Why is freeze-dried food important for space travel? Underline the text in the passage that helped you answer. _____

_____

Day 2

Jon Comer is a professional skateboarder who has attended competitions around the world. He drops into a half-pipe and does many difficult tricks. What is really amazing is that Comer has an artificial limb, called a *prosthetic*, on one leg. Technology and science join to help many people like Comer. The limb is made of a kind of flexible plastic with a microprocessor inside. When joined to the body, muscle movement is changed to electric signals, which makes the artificial limb move.

1. How did science and technology combine to create helpful prosthetics? Underline the text in the passage that helped you answer. _____

_____

Day 3

At the age of nine, Brandon Whale visited a hospital and saw young children crying when they got shots. Whale discovered that the children's veins were small and hard to locate. Also, they were even harder to find when the children were anxious and tense. He developed a soft ball that looked like a beetle. Children could squeeze it, which helped the kids—and their veins—relax. Brandon called his invention the "Needle Beetle." A toy company liked the idea so much that they started to manufacture his invention.

1. Why did Whale invent the "Needle Beetle"? Underline the text in the passage that helped you answer. _____

_____

Day 4

Engineering

Circle the best answer.

1. Which is a characteristic of a good engineer?

   A. observant

   B. happy

   C. quick

   D. selfish

Answer the questions.

2. Is an engineer a scientist? Explain.

   _____

   _____

   _____

3. Engineers use their products to solve problems. Think of a problem you would like to solve. Design an invention for it. Write a description of the invention and tell how it solves the problem.

   _____

   _____

   _____

   _____

   _____

   _____

   _____

   _____

   _____

   _____

   _____

   _____

   _____

4.RI.1, 4.RI.10, 4.W.2, 4.L.6

Healthy Habits

Write three foods that belong in each food group.

1. Grains: _____

2. Vegetables: _____

3. Fruits: _____

4. Dairy: _____

5. Proteins: _____

6. Oils: _____

1. Why is it important to eat a healthy diet? Give two reasons.

   _____

   _____

   _____

2. Why should you eat foods of all different colors?

   _____

   _____

   _____

1. Name three ways that exercise helps the body.

   _____

   _____

   _____

   _____

   _____

1. Read the label. Is this a healthy snack? Explain.

   _____

   _____

   _____

   _____

   _____

Nutrition Facts	
Serving Size: 1 bag • 1.75 oz	
Amount Per Serving	
Calories 280	Calories from Fat 162
Total Fat 18g	28%
Cholesterol 0mg	0%
Sodium 340mg	14%
Total Carbohydrate 25g	8%
Protein 3g	6%

Healthy Habits

Circle the best answer.

1. Which group of foods is an example of a healthy lunch?

    A. an apple, an orange, and juice

    B. turkey sandwich on wheat bread, kiwi, and milk

    C. ham sandwich on white bread, chips, and yogurt

    D. broccoli, carrots, and milk

Answer the questions.

2. Choose an answer from number 1 that is not a healthy lunch. Explain why it is not healthy and how you would change it to be a better choice.

    _____

    _____

    _____

3. As a 10-year-old boy, Nathan needs about 1,800 calories a day. For breakfast, he has 410 calories. For lunch, he has 490 calories. He eats an after-school snack that is 210 calories and then plays soccer with his sister for 30 minutes, which burns 180 calories. He has 525 calories for dinner. Is Nathan's daily calorie total more or less than his daily recommended intake?

4. Are Nathan's habits healthy? Why or why not? Explain to Nathan what he should continue doing and what he could change and why. Discuss your answer with a partner.

    _____

    _____

    _____

    _____

    _____

    _____

    _____

4.RI.7, 4.SL.1, 4.OA.A.3, 4.NBT.B.4

Name_____

Pollution

Write a word to correctly complete each sentence.

1. Adding harmful materials to the environment causes _____.
2. Dumping trash on the ground results in _____ pollution.
3. Construction sites using heavy equipment and loud music produce _____ pollution.
4. Sewage and oil leaking into ponds, lakes, streams, and oceans make _____ pollution.
5. _____ pollution is a big problem because it can affect our breathing and the temperature balance on Earth.

Day 1

1. List three causes of air pollution.

_____

_____

_____

2. How does air pollution affect plants?

_____

_____

_____

Day 2

1. List three causes of water pollution.

_____

_____

_____

2. How does water pollution affect animals?

_____

_____

_____

Day 3

1. Most of what people throw away ends up in landfills. What problem arises when it rains near some landfills?

_____

2. What is one way that a community can reuse a landfill? How does this help the community?

_____

_____

_____

Day 4

Pollution

Look at the graph. Then, answer the questions.

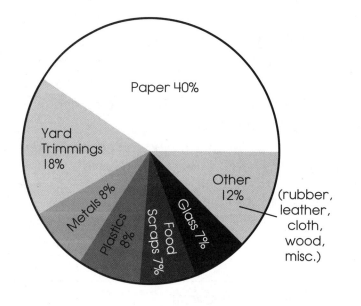

1. Write a caption under the graph.

2. Write the types of trash in order from least amount to the greatest amount.

   _____

   _____

3. Which kind of trash is thrown out the most? Give three examples of this kind of trash that you throw out.

   _____

   _____

4. Your class is having a discussion on reducing waste. Give one example of how you can reduce the waste in three of the above categories. Compare your examples with a partner's.

   _____

   _____

   _____

   _____

4.RI.7, 4.SL.1, 4.L.4, 4.L.6, 4.NBT.A.2

Name_____

Global Warming

1. What is the greenhouse effect?

_____

_____

2. Why is this system important to Earth?

_____

_____

_____

1. What does the picture show?

_____

_____

_____

_____

2. How does this affect nature?

_____

_____

1. What is global warming?

_____

2. List three specific causes of global warming.

_____

_____

_____

_____

1. What is the main effect of global warming?

_____

2. List three ways that the balance of nature is affected by global warming.

_____

_____

_____

_____

Global Warming

Answer the questions.

1. Write notes for an oral report on how the greenhouse effect and global warming are related and give several specific ways people can reduce global warming. Use the box to draw a visual display to enhance your report.

_____

_____

_____

_____

_____

_____

_____

_____

_____

_____

_____

_____

_____

_____

4.RI.7, 4.SL.4, 4.SL.5, 4.L.6

Resources

1. What is a natural resource?

   _____

   _____

2. Write **R** beside each item that is a natural resource.

   _____ air   _____ corn   _____ water

   _____ cow   _____ rock   _____ coal

   _____ shirt   _____ electricity   _____ paper

1. What is a renewable resource? Give two examples.

   _____

   _____

   _____

2. What is an inexhaustible resource? Give two examples.

   _____

   _____

   _____

1. Ivy doesn't understand why nonrenewable resources should be used carefully. Tell how you would respond to her. Share your response with a partner.

   _____

   _____

   _____

   _____

   _____

1. What are three ways that people can help conserve natural resources?

   _____

   _____

   _____

   _____

   _____

Resources

Answer the questions.

1. What is petroleum? Is it a renewable or nonrenewable resource? Why?

   _____

   _____

2. In 2014, a logging company cut an area of forest 78 yards long by 26 yards wide. The forest was 3,375 square yards. How much area of the forest is left? If it takes 120 years for harvestable trees to regrow, in what year will that part of the forest be ready to cut again?

3. Your friend argues that petroleum should not be conserved. How would you respond?

   _____

   _____

   _____

   _____

   _____

Circle the best answer.

4. What kind of resource is water?

   A. exhaustible

   B. inexhaustible

   C. renewable

   D. nonrenewable

5. How does a lumber company make sure that trees are a renewable resource?

   A. They don't cut too many trees.

   B. They make fewer paper products.

   C. They plant trees in places they log.

   D. They only cut in rain forests.

4.SL.1, 4.L.6, 4.OA.A.3, 4.NBT.B.5, 4.MD.A.3

Conservation

Unscramble the letters in parentheses to complete each sentence.

1. People practice _____ (rasvencoonit) when they protect and use natural resources wisely.
2. Some animal populations, like the dinosaurs, are _____ (cixtent).
3. Other populations are _____ (gadneenerd), meaning there are few of them left.
4. Many people work together to _____ (torcept) these organisms.
5. They put aside land as a _____ (furege) where the plants and animals can live in safety.

1. Circle the picture that shows the plowing method that will best conserve the soil.

2. Write a caption explaining why a farmer would do this.

_____
_____
_____

1. Why are forests important? Name two reasons.

_____
_____

2. In many forests, lumber companies are supposed to plant new trees to replace the ones they cut. Why?

_____
_____
_____

1. Why are some animals endangered? Name two reasons.

_____
_____

2. Why do people protect endangered animals?

_____
_____
_____

Name_____

Conservation

Answer the question.

1. More than half of all plants and animal species in the world live in the rain forest. Scientists believe that millions more exist but have not yet been discovered. However, in the last 50 years, nearly half of the rain forests have been destroyed. Write a short report explaining why the rain forest ecosystem is important and should be saved.

_____

_____

_____

_____

_____

_____

_____

_____

_____

_____

_____

_____

_____

_____

_____

_____

_____

_____

_____

_____

_____

4.RI.7, 4.W.1, 4.W.10, 4.L.4, 4.L.6

Page 9

Day 1: 1. ruler; 2. thermometer; 3. balance scale; 4. graduated cylinder; 5. clock or stopwatch; **Day 2:** 1. A hand lens makes things look larger than they are. Examples will vary.

Day 3: 1. Make sure the pointer on the base sits on zero when the pans are empty. Place the object you are measuring in the left pan. Add the mass weights one at a time, from largest to smallest, to the right pan to get the pointer back to zero. Add the mass weights to find the total grams of the object.

Day 4: 1. A microscope makes objects that cannot be seen with the eyes look much bigger and closer. Examples will vary. 2. A telescope makes objects that are very far away look bigger and closer. Examples will vary. 3. Binoculars make objects that are at a medium distance look bigger and closer. Examples will vary.

Page 10

1. C; 2. D; 3–4. Answers will vary.

Page 11

Day 1: 1. The customary system uses feet to measure length, pounds to measure weight, cups to measure capacity, and Fahrenheit degrees to measure temperature. 2. The metric system is based on tens. It uses meters to measure length, grams to measure mass, liters to measure volume, and Celsius degrees to measure temperature. **Day 2:** 1. the metric system; 2. The metric system allows scientists around the world to understand the data gathered and repeat experiments, even if they do not speak the same language. **Day 3:** 1. L; 2. g; 3. m; 4. cm; 5. mL; 6. kg; 7. mm; 8. km; **Day 4:** 1. 1,000; 2. 100; 3. 1,000; 4. 0.01; 5. 1,000; 6. 10; 7. 100; 8. 1,000

Page 12

1. A; 2. B; 3. Answers will vary but may include a crayon. 4. Answers will vary but may include the height of a desk. 5. Answers will vary but may include a bottle of water. 6. Answers will vary but may include a can of soup. 7. Answers will vary but may include a paper clip. 8. Wesley jumped farther. His jump of 1,000 centimeters is equal to 10 meters, which is farther than 9 meters.

Page 13

Day 1: 1. observing, using your five senses to learn about the world; 2. classifying, grouping objects based on characteristics or qualities; 3. communicating, sharing information using words and visual aids; 4. inferring, using what you know and what you learn to make conclusions; 5. predicting, making an educated guess about what will happen; 6. comparing, telling how objects are alike and different; **Day 2:** 1–2. Answers will vary. **Day 3:** 1. A variable is something that can be changed in an experiment. 2. Identifying and changing variables allows a scientist to more thoroughly and accurately complete an experiment. **Day 4:** Answers will vary but may include so that scientists can repeat the experiment and can share accurate data.

Page 14

1.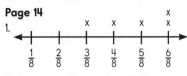

2. Answers will vary. 3. A; 4. D

Page 15

Day 1: 1. 6, 7, 3, 1, 4, 5, 8, 2; **Day 2:** 1. A hypothesis is a statement predicting the result of an experiment or activity. 2. B; **Day 3:** 1. Answers will vary but may include that he could use materials that have different textures, colors, or surface finishes, or different kinds of light. 2. Jay will change one variable each time. If he changes two or more at a time, then he will not know which variable caused the end result. **Day 4:** 1. true; 2. false; 3. false; 4. false; 5. true

Page 16

1. Answers will vary but may include that the scientific method helps the scientists stay organized. Other scientists need to be able to do the experiment and get the same results to show the conclusion is correct. 2. Answers will vary but should include that Drew should state a new hypothesis and do another experiment. 3. Answers will vary but may include that communicating information helps everyone learn new facts about the world. Other scientists can use the information as a foundation for a new experiment. 4. D; 5. A

Page 17

Day 1: 1. Answers will vary. **Day 2:** 1. Answers will vary but may include checking the methods used and repeating the experiment. **Day 3:** 1. hypothesis: a proposed explanation that needs to be tested; 2. theory: a scientifically accepted explanation that has been tested repeatedly; 3. law: a scientifically accepted principle that does not explain a phenomenon; **Day 4:** 1. Answers will vary but may include so that other scientists can repeat the experiment, the conclusion is considered valid, or other scientists can alter the experiment to test another hypothesis.

Page 18

1. Answers will vary.

Page 19

Day 1: 1. an objective detail about something; 2. a conclusion based on observations; **Day 2:** 1. Answers will vary. **Day 3:** 1. O; 2. I; 3. O; 4. O; 5. I; **Day 4:** 1. Answers will vary but may include magnets do not attract wood. 2. Answers will vary but may include talc and chalk are softer than quartz. 3. Answers will vary but may include sharp beaks help predatory birds tear their food.

Page 20

1–4. Answers will vary.

Page 21

Day 1: 1. anything that has mass and occupies space; 2. Answers will vary but may include an orange is a solid on the outside, has some liquid on the inside, is round, or is orange in color. **Day 2:** 1. A potato has more mass because it has more matter in it. 2. fish tank (circled); A fish tank has more volume because it can hold more water. **Day 3:** The three states of matter are solid, liquid, and gas. Examples will vary. 2. The pictures show that a liquid can take the shape of its container. **Day 4:** 1. L, G; 2. G; 3. S; 4. S, L, G; 5. L

Page 22

1. The cup is filled with air, so water cannot enter a space that is already filled with matter. 2. No. One object may have more matter in it, so it will have more mass. 3. Answers will vary but may include hair, stomach acid, and oxygen. 4. Check students' drawings. The solid particles should be close together and in a regular pattern. The liquid particles should be farther apart. The gas particles should be the farthest apart.

Page 23

Day 1: 1. In a physical change, the substance stays the same. Examples will vary. 2. In a chemical change, the substance becomes a different substance. Examples will vary. **Day 2:** 1. P; 2. C; 3. P; 4. C; 5. P; 6. C; 7. P; 8. P; 9. C; 10. P; **Day 3:** 1. Answers will vary but should include an explanation of a solution. **Day 4:** 1. Cup B's mixture is more concentrated because it has more powder mixed in.

Page 24

1. A, C, B; $\frac{7}{10}$ L more of the substance; 2. Answers will vary but should include that a solution is a type of mixture where a substance is dissolved into another substance. 3. In a physical change, the substance stays the same. In a chemical change, the original substance changes to a different substance. Examples will vary. 4. D; 5. A

Page 25

Day 1: 1. the ability to do work; 2. Answers will vary. **Day 2:** 1. the ability to do work; 2. the energy an object has because of its position; 3. the energy of an object because it is moving; 4. the energy of motion; 5. the energy caused by the flow of electricity; 6. the energy caused by chemical change; **Day 3:** Students should label the top of the hill **P**, and the bottom of the hill **K**. **Day 4:** 1. false; 2. true; 3. true; 4. false; 5. true; 6. false

Page 26

1. Answers will vary. 2. Answers will vary but may include that radiant energy travels in waves and that it can travel through space where there is no matter. 3. The skier changes potential energy of standing at the top of the hill into kinetic energy by pushing off and moving down the hill. 4. Check students' drawings. 5. B

Page 27

Day 1: 1. No. Energy can only change form. **Day 2:** 1. a car burning gasoline to move; 2. turning on a lamp; 3. using an electric hot plate; 4. rubbing hands together to warm them; 5. using batteries to power a laptop; **Day 3:** 1. Energy changes from electrical to radiant. 2. Energy changes from chemical to mechanical and thermal. **Day 4:** 20 times

Page 28

1–2. Answers will vary. 3. 111 feet; 4. C

Page 29

Day 1: 1. A force is a push or a pull. 2. Answers will vary but may include force can cause an object to start moving, stop, change direction, slow down, or increase speed. **Day 2:** 1. gravity; 2. friction; 3. inertia; 4. The force of gravity is helping to pull a person down the stairs. A person walking up the stairs is going against the force of gravity, so it will be a harder task. **Day 3:** 1. false; 2. true; 3. false; 4. true; 5. true; **Day 4:** 30 N

Page 30

1. The floor is pushing back on Jayla's hands with an equal force. 2. The force of Dion's hands stops the ball. 3. The force of friction of the brakes slows Sam's bike. 4. The force of gravity pulls the roller coaster down. 5. The force of the bat changes the direction of the ball. 6. B

Page 31

Day 1: 1. An electron has a negative charge. A proton has a positive charge. A neutron has no charge. 2. An electric charge is created when electrons transfer from one object to another so that one object has more negative charges. **Day 2:** 1. battery or energy source; 2. wire; 3. switch; 4. bulb; 5. The diagram shows a closed circuit. **Day 3:** 1. Electrons leave the battery through the negative end. The flow creates an electric charge that travels through the wire to the bulb, which lights. The current continues to flow through the next wire to the switch. From there it moves through the third wire and back in the positive end of the battery. 2. If the switch is open, the circuit is broken and the current cannot flow throughout the system. **Day 4:** 1. The inside of the wire is metal. Metal is a good conductor and can easily move an electric current. People who touch live wires can get shocked. Plastic is an insulator and stops the flow of an electric current. People can hold the plastic-coated wires and not get shocked.

Page 32

1. D; 2. B; 3. The light goes out because turning off the lamp breaks the flow of electricity through the circuit. 4. The string of lights is a parallel circuit. Each bulb has its own path for electricity. One bulb can burn out and not stop the flow of the electric current to the rest of the bulbs. 5. Answers will vary.

Page 33
Day 1: 1. A magnet is an object that will attract iron, steel, and other certain metals. 2. The ends, or poles, of the magnet have the greatest force. **Day 2:** 1. poles; 2. south; 3. north; 4. attract; 5. repel; **Day 3:** 1. From top to bottom, the poles should be labeled **N, S, S, N** or **S, N, N, S.** 2. The same poles of the magnets face each other. Because they are like poles, the magnetic fields repel each other. **Day 4:** 1. true; 2. false; 3. true; 4. true; 5. false

Page 34
1. Answers will vary but may include wood, plastic, and water. 2. The needle of the compass is magnetized and is attracted to Earth's magnetic north pole. 3. Check students' drawings. 4. Seth is incorrect. Check students' reasoning.

Page 35
Day 1: 1. Electricity is caused by the flow of electrons, which have a negative charge. Their charge creates a magnetic field that can be turned on and off with the flow of electricity. **Day 2:** 1. door locks; 2. hard drives; 3. generators; 4. cranes; **Day 3:** 1. Check students' drawings. **Day 4:** The two like poles are facing each other. Because they repel, the maglev rises into the air and moves.

Page 36
1. Electricity moves through a wire that is wrapped around iron, which creates a strong magnetic field. 2. The doorbell does not need energy all of the time. 3. Answers will vary but may include he was observant to notice that the compass needle moved, and he shared his knowledge with others. 4. The system needs a source of energy, wire, and iron. Check students' underlining.

Page 37
Day 1: 1. Answers will vary. 2. All sounds are made when things vibrate. **Day 2:** 1. Sound waves are the movement of molecules as they vibrate. 2. Sound waves move in a straight line. When the vibrations move through matter, they push close together. When they pass, they move farther apart. **Day 3:** 1. Volume is how loud or soft a sound is. 2. A loud sound takes more energy to produce. The sound wave is bigger and travels farther. **Day 4:** 1. Students should circle the sound wave with crests and troughs that are farther apart. 2. Answers will vary but may include the vibration in something with a low pitch will be slow. The wavelength will be more spread out.

Page 38
1. Sound travels through all states of matter. 2. aluminum; 3. Answers will vary but may include aluminum is a solid. The molecules in a solid are packed more tightly, so they can pass the sound more quickly. 4. Answers will vary but may include air because you want to stop or slow sound. Since sound travels more slowly in air, it would be the best material listed in the graph to use.

5.

Material	Distance in 1 second* (in m)	Distance in 1 minute (in m)
Air	340	**20,400**
Water	1,490	**89,400**
Aluminum	5,100	**306,000**
Wood	3,850	**231,000**

Page 39
Day 1: 1. Visible light is only a small part of the electromagnetic spectrum. There are many other light waves that are not visible to humans. **Day 2:** wavelengths, Ultraviolet, microwaves, Radio, X-rays, gamma rays; **Day 3:** Answers will vary. **Day 4:** 300,000; 1. 4,474 feet

Page 40
1. Light travels in waves and in straight lines. 2. Check students' drawings. Reflected light should bounce back at an equal angle. Refracted light should bend before continuing in a straight line. Absorbed light should end. 3. Colors are certain wavelengths of visible light. The color of an object is reflected back at the eye, while all other wavelengths are absorbed by the object. 4. D; 5. A

Page 41
Day 1: 1. Students should write a **P** in front of can grow, made of cells, makes its own food, can reproduce, cannot move; **Day 2:** 1. One group of plants makes seeds. Examples will vary. One group of plants does not make seeds. Examples will vary. 2. Yes, a daisy and a pine tree belong in the same main group because both plants make seeds. **Day 3:** 1. stamen: makes the pollen for fertilization; 2. sepal: covers the flower buds; 3. petal: keeps the parts that make seeds safe; 4. pistil: makes the eggs that grow into seeds; **Day 4:** 1. Moss reproduces by small cells called spores. 2. Pollination is the process where pollen moves from the stamen to the pistil. The wind, insects, and other animals help carry the pollen from one part of the flower to the other.

Page 42
1. After a flower is fertilized with pollen, each pollen grain grows a tube from the pistil down to the ovary. Sperm in the pollen joins with an egg in the ovary to make a seed. 2. Insects visit the flowers to get nectar. The pollen sticks to their bodies. As they visit other flowers, the pollen sticks to the pistils. 3. Wind or insects carry the pollen from the top of the plant to the flowers below to pollinate them.
4.

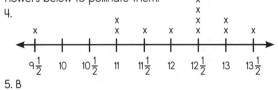

5. B

Page 43
Day 1: 1. reproduce; 2. food; 3. shelter; 4. young; 5. oxygen;
Day 2: 1. Vertebrates have backbones. Examples will vary. Invertebrates do not have backbones. Examples will vary. 2. Answers will vary but may include they are alike because they do not have backbones. They are different because the beetle has an exoskeleton and jointed legs. The worm has a soft body and no legs. **Day 3:** 1. lives in freshwater or salt water; 2. lives part of its life in the water and part of it on land; 3. has a body covered in dry scales; 4. has a body covered in feathers; 5. feeds its young milk; **Day 4:** 1. An exoskeleton is the hard outer shell that covers the body of some animals. It protects the soft inside parts. 2. If an animal is cold-blooded, the animal's body temperature changes with the temperature of its surroundings. Fish, reptiles, and amphibians are cold-blooded.

Page 44
1. The backbone is made up of many small bones and runs along the middle of an animal's back. It supports the body. It is an important characteristic because scientists use it to classify animals into two main groups—those with backbones and those without. 2. Fish: bass, shark; Reptile: snake, alligator; Amphibian: salamander, frog; Bird: hawk, parrot; Mammal: whale, human; 3. Answers will vary but may include a butterfly does not have a backbone, but a bird does. 4. D

Page 45
Day 1: 1. heredity; 2. diversity; 3. species; 4. trait; 5. inherit;
Day 2: Answers will vary but may include they may have a better chance of survival, or could reproduce more and pass on those stronger traits. 2. Answers will vary but may include an animal may be more noticeable and become easy prey, or it may not be accepted by the species and be left to die.
Day 3: 1. Answers will vary but may include they need to inherit traits so that they can survive in the habitat. 2. Answers will vary. **Day 4:** 1. Answers will vary.

Page 46
1. Heredity makes sure that certain traits are passed down in a species so that they can survive in the habitat. 2. There are specific traits inherited from a parent, such as eye color. There are inherited traits that are affected by the body systems and environment, such as height. There are acquired skills, like playing an instrument. 3. Answers will vary.

Page 47
Day 1: 1. a change in an organism that makes it better suited to survive in its environment; 2. Answers will vary. **Day 2:** 1. S; 2. S; 3. B; 4. B; 5. B; 6. S; **Day 3:** 1. Answers will vary but may include the animal's large ears help it hear predators better, it has adapted to living underground, its short fur keeps it cool, and it has learned to stay hidden in the entrance of its home to check for predators. **Day 4:** 1. a physical change in an organism; 2. Answers will vary but may include that the

mutation can help it better survive in its environment, but it can also make the animal stand out more to predators.

Page 48
1. Check students' drawings and captions. 2. Answers will vary. 3. Answers will vary. 4. Answers will vary but may include the bird's long legs help it stand in deep water, and its long, pointy beak helps it catch fish.

Page 49
Day 1: 1. A behavior is the way a living thing acts. 2. Answers will vary but may include a dog wags its tail, a fish swims, an ant digs tunnels underground, and a human laughs.
Day 2: 1. L; 2. I; 3. I; 4. I; 5. L; 6. L; **Day 3:** 1. A reflex is an automatic behavior controlled by nerves. Examples may include blinking or breathing. 2. A stimulus is an action or event that causes a behavior. The response is the behavior to the stimulus. Examples will vary but may include a ringing phone is a stimulus and the response is answering it. **Day 4:** 1. Animals learn behaviors through training. Examples will vary but may include people train dogs to sit using a stimulus, such as food. They also learn behaviors by watching. Examples will vary but may include lion cubs learn to hunt by watching adults.

Page 50
1. A; 2. Answers may vary. Instincts are inborn behaviors that offspring inherit. It helps the animal survive and reproduce. Examples will vary but may include instinct drives squirrels to gather and store nuts for the winter months. Instinct also helps birds build nests. 3. Check students' writing.

Page 51
Day 1: 1. photosynthesis; Plants need sunlight, water, and carbon dioxide. 2. Chlorophyll in the leaf traps light energy. Carbon dioxide is taken in through small holes on the underside of the plant. Carbon dioxide and water react together in the presence of chlorophyll and light to make sugar.
Day 2: 1. energy; 2. sugar; 3. herbivore; 4. carnivore; 5. omnivore; **Day 3:** 1. Some animals, like buzzards, eat the dead animals to get the energy. Decomposers, like worms and mushrooms, break down the bodies of dead organisms. 2. A decomposer returns the nutrients stored in dead organisms to the soil so that plants can use them.
Day 4: 1. 4, 3, 1, 6, 2, 5

Page 52
1. Answers will vary. 2. Answers will vary but may include human building can destroy habitats that affect parts of a food chain. 3. decomposer: an organism that breaks down the bodies of dead organisms for energy; photosynthesis: the process by which plants turn sunlight into energy; food web: interconnected food chains; consumer: an organism that eats other organisms for energy; producer: an organism that produces energy; 4. A; 5. B

Page 53

Day 1: 1. An ecosystem is made up of living and nonliving things in an environment. 2. Answers will vary but may include forest, pond, mud puddle, or backyard. 3. Answers will vary.
Day 2: 1. Answers will vary. **Day 3:** 1. Answers will vary but may include flood, fire, drought, disease, or the building of roads, buildings, or farms. 2. Answers will vary but may include plants and animals will adapt, move, or die. **Day 4:** 1. the place an animal lives that provides for all of its needs, including water, food, and shelter; 2. An ecosystem includes many different living and nonliving things existing together in a place. A habitat is more specific to an animal's needs for survival.

Page 54

1. Answers will vary but may include people are living in the foxes' habitat. The foxes may not be able to have their food needs met, so they are going into the garbage cans to find food to eat. 2. Answers will vary.

3.
Animal	Population in 2014	Population in 2015
squirrels	1,536	512
cardinals	642	214
robins	903	301
rabbits	594	198
chipmunks	708	236

4. Answers will vary.

Page 55

Day 1: 1. a group of events that happen in the same order over and over; 2. life; 3. oxygen; 4. photosynthesis;
Day 2: 1. Bean Seed Growth; 3, 1, 4, 5, 2; **Day 3:** An adult lays an egg on a milkweed plant. The egg hatches into a larva. The larva, also known as a caterpillar, eats the milkweed and grows. It attaches itself to a leaf and becomes a chrysalis. After metamorphosis, the pupa hatches as an adult butterfly.
Day 4: 1. Answers will vary but may include the life cycles of organisms are often related to seasons because weather is a factor in the growth of organisms. Many baby animals born each spring would not be able to survive during harsh winters or heated summers. However, plants, like apples, need the cold in order to produce.

Page 56

1. B; 2. A; 3. A frog lays thousands of eggs. The eggs hatch into tadpoles. The tadpoles grow back legs and become froglets. The froglets grow front arms, and their tails shrink. When its tail disappears, the froglet is an adult frog. 4. Answers will vary but may include energy flows through a food chain or web in a cycle. Organisms rely on cycles for energy and growth.

Page 57

Day 1: 1. all the living and nonliving things in a place; 2. the place an animal lives where all of its needs can be met; 3. a group of one kind of living thing that lives in a place; 4. everything that is around a living thing; 5. all of the

populations that live in a place; **Day 2:** 1. Scientists want to find out if a population is changing. If the number of organisms is getting smaller, they may want to protect them. If the numbers are getting larger, they may want to reduce the number to maintain the balance of nature. 2. Answers will vary but may include they put bands or transmitters on the animals.
Day 3: 1. The rabbit population grew. 2. It grew from the first to eighth year. **Day 4:** Answers will vary but may include when the rabbit population grew, so did the coyote population. When it got smaller, so did the coyote population. Since coyotes eat rabbits, when there were fewer rabbits to eat, the coyote population also changed due to lack of food.

Page 58

1. Answers will vary. 2. Answers will vary but may include the population of plants will grow because the mice are not there to eat them. The population of snakes may get smaller without mice to eat unless they are able to change their diet. The hawks' population may shrink if the number of snakes shrinks.

3.
Year	0	1	2	3	4	5
Owls	208	416	832	1,664	1,456	1,274
Mice	532	1,064	2,128	4,256	3,724	3,259

4. D

Page 59

Day 1: 1. revolution; 2. rotation; 3. third; 4. star; 5. axis;
Day 2: 1. Check students' labeling. 2. The diagram shows how Earth revolves around the sun, the moon revolves around Earth, and the moon and Earth both rotate. **Day 3:** 1. elliptical; 2. Venus, Earth, Saturn, Mercury; **Day 4:** 1. Answers will vary but should include an explanation of how Earth's tilted axis creates seasons that differ across the northern and southern hemispheres.

Page 60

1. A. winter, summer; B. spring, autumn; C. summer, winter; D. autumn, spring; 2. Answers will vary but should include an explanation of the differences in the orbits of Mars and Earth. 3. Check students' angles. 4. true; 5. false; 6. true

Page 61

Day 1: 1. true; 2. false, It takes exactly $29\frac{1}{2}$ days for the moon to circle around Earth. 3. false, The moon reflects the sun's light. 4. true; **Day 2:** 1. 4, The growing moon's surface is mostly lit. 2. 7, One-half of the shrinking moon is lit. 3. 6, The shrinking moon's surface is mostly lit. 4. 2, A sliver of the growing moon is lit. 5. 3, One-half of the growing moon is lit. 6. 8, A sliver of the shrinking moon is lit. 7. 1, The moon looks dark in the night sky. 8. 5, The moon looks like a big, bright circle. **Day 3:** 1. Students should draw a mostly lit moon near Earth near the 11:00 position. **Day 4:** 1. a solar eclipse; 2. In a solar eclipse, the moon is between Earth and the sun. It makes a shadow on Earth during the day.

Page 62

1. In a solar eclipse, the moon is between Earth and the sun and makes a shadow on Earth in the daytime. In a lunar eclipse, Earth is between the sun and moon, and Earth's shadow darkens the moon at night. 2. The new moon is the beginning of the phases of the moon. The dark half faces Earth, so it cannot be seen.

3. 29

Full Moon	Date	Full Moon	Date
1	January 18	7	July 11
2	February 16	8	August 9
3	March 17	9	September 7
4	April 15	10	October 6
5	May 14	11	November 4
6	June 12	12	December 3

4. third-quarter moon; waning crescent moon

Page 63

Day 1: 1. A mineral is nonliving solid matter found in nature. 2. The following phrases should have a **P**: how hard it is; how shiny it is; how it breaks; its color; if it is magnetic;
Day 2: 1. Sedimentary rocks are formed when many layers of sediment pile on top of each other. They are pressed together and harden. 2. Igneous rocks are formed when hot, liquid rock cools and hardens. 3. Metamorphic rocks are formed from sedimentary or igneous rock that has been heated under pressure. **Day 3:** 1. fossil; 2. sedimentary; 3. imprint; 4. coal; 5. igneous; **Day 4:** 1. Soil is made of tiny rocks, air, water, and humus (dead plants and animals). 2. Over thousands of years, rocks are weathered by water, wind, plants, and ice, which break them into tiny pieces. Decomposers eat dead plants and animals, and their waste, full of nutrients and energy, mixes with the weathered rocks.

Page 64

1. Answers will vary but should include that rocks may be melted, cooled, heated under pressure, weathered, eroded, or compacted. Depending on the beginning type of rock and the change the rock goes through, it will become sediment, magma, or one of the three types of rock. 2. Answers will vary. 3. C; 4. A

Page 65

Day 1: 1. mold; 2. dinosaur; 3. amber; 4. paleontologists; 5. cast; **Day 2:** 1. 3, 2, 1, 5, 4; **Day 3:** 1. Scientists study fossils to learn about organisms that lived long ago, how Earth has changed, and how organisms have adapted to changing environments.

2. Answers will vary but may include scientists can compare the fossils they find with the animals living today. If the parts are the same, they can make inferences and draw conclusions that the animals looked and acted similarly. **Day 4:** 1. a sedimentary or limestone rock; 2. Most fossils are found in sedimentary rock. The animal died and was covered in mud that hardened through pressure.

Page 66

1. Answers will vary but may include fossils' locations give clues about the shape of land, the climate, and the time period in which the organism lived. Sedimentary rock filled with marine fossils indicates that water covered the area at one time. Coal, made from ferns that grow in wet climates, indicates the area had a wet climate at one time. The rock layer in which the fossil is found shows an approximate time it lived. 2. 4, 1, 2, 3; 3. She can infer that Antarctica was a warm, wet place long ago. Ferns only grow in this kind of climate; finding a rock with that print shows the plant lived there, so it must have been warm. 4. A

Page 67

Day 1: 1. Answers will vary but may include the lava from a volcano burns the land as it flows. It also builds new land when the lava flows into the cold ocean water and hardens. 2. Answers will vary but may include an earthquake can make large cracks in Earth. 3. Answers will vary but may include a flood can wash away buildings and roads, as well as move dirt and rocks to new places, reshaping the land. **Day 2:** 1. C; 2. P; 3. C; 4. P; 5. P; 6. C; **Day 3:** 1. Weathering is the process where rocks are broken down into smaller pieces. Erosion is moving rocks and soil to new places. 2. water, wind, and ice; 3. As a glacier moves across the land, its weight pushes rocks and soil. **Day 4:** 1. Deposition; 2. slows; 3. nutrients; 4. dunes; 5. beaches; 6. delta

Page 68

1. Rocks are weathered by water, wind, ice, and plants. The flow of water chips off tiny grains and pieces. Wind carries pieces of sand that hit bigger rocks, causing them to chip. Water also gets in cracks and freezes. As the ice expands, it forces the cracks to grow and can break the rock. Plant roots grow in the cracks of rocks. As they grow, they break the rocks. 2. 60–80 years; 3. Answers will vary but may include weathering from ice, wind, water, and plants can break off a piece. Erosion from the flow of rain can force it to move across the land into a river. The river continues to move the piece to the end of the river. When the water slows, deposition allows the rock to settle in a delta. 4. C

Page 69

Day 1: 1. volcano; 2. eruption; 3. lava; 4. ash; 5. dormant;

Day 2: 1. An earthquake is caused when the plates of the earth move, which makes the ground shake along a fault.
2. Answers will vary but may include an earthquake creates cracks in the earth. It causes human-made structures to fall.

Day 3: 1. Answers will vary but may include floods can erode the land, taking away soil, trees, and human-made structures. Floods can even reshape the land and the flow of a river.
2. Answers will vary but may include some floods leave behind rich soil that is good for growing crops. **Day 4:** Both are dangerous storms with high wind speeds that can cause a lot of damage. A tornado is a turbulent windstorm in which a funnel-shaped wind drops from a thundercloud and moves quickly in a narrow path across the land. It may last a few minutes or as long as an hour. A hurricane begins over the ocean. High winds and heavy rains move slowly across the water in a wide band and may strike land. A hurricane can last for more than two weeks.

Page 70

1. No, the land will be different due to changes made by events caused in nature. 2. Answers will vary but may include a hurricane is easier to predict since forecasters can track the movement for a long time before it strikes land. A tornado happens in thunderstorm conditions, which can appear suddenly. 3. Answers will vary but may include Hawaii will get bigger because the lava is spreading out and forming new land. 4. Category 4; 5. Category 3; 6. between 20 and 25 minutes

Page 71

Day 1: 1. Air temperature rises because energy from the sun heats the land and water. The air above them is then heated, raising the temperature. 2. Land absorbs the sun's energy more quickly, while water absorbs it more slowly. **Day 2:** 1. As air near the surface of earth heats, it gets lighter and rises. The cold air higher up is heavier and falls. The falling air heats as it gets near the surface and begins to rise. The cycle causes wind. 2. the amount of water vapor in the air; **Day 3:** 1. Clouds form when the sun's energy heats water. The water evaporates and becomes water vapor that rises with warm air. The air cools, and the vapor condenses, changing back to tiny, liquid-water droplets. The droplets form clouds. 2. Answers will vary but should include that the clouds are cumulus clouds and signal good weather, so Kim should have a picnic. **Day 4:** 1. The picture shows a warm front. A warm air mass is lighter and climbs over the heavier, colder air mass. The rising air cools and forms rain clouds.

Page 72

1. Answers will vary.

2.

Type of Cloud	Typical Altitude
stratus	0–**6,500** ft.
stratocumulus	2,000–**6,500** ft.
cumulus	2,000–**3,000** ft.
cumulonimbus	2,000–**45,000** ft.
altocumulus	**6,500–18,000** ft.
altostratus	6,500–**16,500** ft.
nimbostratus	2,000–18,000 ft.
cirrocumulus	**16,500–45,000** ft.
cirrus	16,500–**45,000** ft.
cirrostratus	20,000–**42,000** ft.

Page 73

Day 1: 1. a group of events that happen in the same order over and over; 2. The following things should have a **C**: a day, the seasons, the rain, the soil, tides; **Day 2:** 1. water, evaporates, life, webs, climate, erode, sediment, rock; **Day 3:** 1. The sun heats water on Earth and changes it to water vapor through evaporation. The water vapor rises into the air and cools as it rises, through condensation. It then changes back into tiny drops of water, which form clouds. The drops become heavy and fall to Earth as precipitation. 2. The water cycle is important because it brings fresh water that organisms need to survive. **Day 4:** 1. false; 2. true; 3. false; 4. true; 5. true; 6. false

Page 74

1. C; 2. C; 3. Answers will vary.

Page 75

Day 1: 1. the application of scientific knowledge to help people in many fields; 2. Answers will vary. **Day 2:** 1. Check students' paragraphs. **Day 3:** Answers will vary but may include that phones used to be connected by wires. Phones have gotten smaller, portable, and capable of more functions.
Day 4: 1. Answers will vary but may include cable is not available in some areas, or the people choose not to pay for it.

Page 76

1. Answers will vary.

Page 77

Day 1: 1. Answers will vary but may include both engineers and scientists try to solve problems and follow a process. Engineers design things to solve problems, while scientists answer questions. Check students' underlining. **Day 2:** 1. Answers will vary but may include that it makes it possible for astronauts to pack less weight and stay in space longer. Check students' underlining. **Day 3:** 1. Answers will vary but may include they allow prosthetics to respond more like normal limbs. Check students' underlining. **Day 4:** Whale wanted to make finding veins in children easier for doctors and for the children. Check students' underlining.

Page 78
1. A; 2. No. Engineers solve problems, and scientists try to explain patterns and behavior. 3. Answers will vary.

Page 79
Day 1: Answers will vary. 1. bread, cereal, pasta; 2. spinach, lettuce, peas; 3. apple, kiwi, orange; 4. yogurt, cheese, milk; 5. chicken, fish, beans; 6. nuts, avocados, olives; **Day 2:** 1. A healthy diet provides nutrients to give the body energy and helps prevent injury and illness. 2. Each color of food has different minerals and nutrients. Eating many colors makes sure that the body gets a variety for better nutrition.
Day 3: 1. Answers will vary but may include exercise keeps the muscles strong, strengthens the heart and lungs, reduces stress, and helps people sleep better. **Day 4:** Answers will vary but should include it is not a healthy snack because it contains lots of trans fats, sodium, and salt, and it has too many calories for a snack.

Page 80
1. B; 2. Answers will vary. 3. less; 4. Answers will vary but may include that he needs more calories each day because he is about 350 calories under his daily-recommended amount. He could eat a larger breakfast each day or have another snack.

Page 81
Day 1: 1. pollution; 2. land; 3. noise; 4. water; 5. air;
Day 2: 1. Answers will vary but may include car fumes, coal-burning electric plants, forest fires, and volcanoes. 2. Answers will vary but may include ash and dust cover the leaves making it difficult for the plants to get the sunlight they need for photosynthesis. **Day 3:** 1. Answers will vary but may include leaking oil tankers, leaking sewage plants, boat fumes, and trash in the water. 2. Answers will vary but may include animals get sick when they drink the polluted water. **Day 4:** 1. Answers will vary but may include there may be chemicals in the landfill, which run off onto the nearby farmland. The poison kills crops or is absorbed into the plants, making organisms that eat them sick. 2. Answers will vary but may include covering the landfill and building parks and office buildings on top, in order to make the land useful again.

Page 82
1. Answers will vary. 2. food scraps/glass, plastics/metals, other, yard trimmings, paper; 3. paper; Answers will vary but may include magazines, homework paper, and craft scraps.
4. Answers will vary.

Page 83
Day 1: 1. The greenhouse effect is the process in which Earth's atmosphere traps some of the sun's rays and heats the land. 2. It is important because it heats Earth enough for organisms to live. **Day 2:** 1. The picture shows that some of the sun's heat is reflected off the ground. It strikes the heavy layer of gases and reflects back to Earth, where it is absorbed, instead of escaping into space. 2. The result is more heat and increased temperatures. **Day 3:** 1. Global warming is the term used to describe the increased temperatures on land due to human actions. 2. Answers will vary but may include burning fossil fuels for heating and powering electric plants, cutting forests, and cow and sheep farming produce gases that result in the global warming. **Day 4:** 1. The main effect is increased temperatures. 2. Answers will vary but may include the balance of nature changes because the storms are more severe, the polar ice caps are melting, ocean levels are rising, and the habitats of many organisms are changing.

Page 84
1. Answers will vary. Check students' drawings.

Page 85
Day 1: 1. A natural resource is something found in nature that is useful. 2. The following items should have an **R**: air, cow, corn, rock, water, coal; **Day 2:** 1. A renewable resource is a resource that can be replaced in a person's lifetime. Examples will vary but may include trees, crops, or animals. 2. An inexhaustible resource is a resource that can be used repeatedly and not be used up. Examples will vary but may include sun, air, or water. **Day 3:** 1. Answers will vary but should include that nonrenewable resources should be used carefully because once they are used, they are gone and cannot be replaced. **Day 4:** 1. Answers will vary.

Page 86
1. Petroleum is a natural resource that was formed from small organisms that lived long ago. It is a nonrenewable resource because there is a limited amount. No more can be created. 2. 1,347 square yards; 2134; 3. Answers will vary but should include that there is a limited amount of petroleum. When it is gone, it cannot be replaced. 4. B; 5. C

Page 87
Day 1: 1. conservation; 2. extinct; 3. endangered; 4. protect; 5. refuge; **Day 2:** 1. Students should circle the picture on the right. 2. Contour plowing will prevent water flow and soil loss by erosion. **Day 3:** 1. Answers will vary but may include forests are important because they are the habitat for many organisms and the photosynthesis of the trees provides oxygen. 2. Trees are natural resources that provide many products. Replaced trees can grow again and be used.
Day 4: 1. Answers will vary but may include they are endangered due to large numbers being killed or changes in the environment. 2. Answers will vary but may include all animals are important in the ecosystem they live in. If one population changes, it often affects the others too.

Page 88
1. Answers will vary.